WAR X

Human Extensions in Battlespace

War X

Human Extensions
in Battlespace

TIM BLACKMORE

UNIVERSITY OF TORONTO PRESS
Toronto Buffalo London

© University of Toronto Press Incorporated 2005
Toronto Buffalo London
Printed in Canada

ISBN 0-8020-8791-4

∞

Printed on acid-free paper

Library and Archives Canada Cataloguing in Publication

Blackmore, Tim, 1961–
    War X: human extensions in battlespace / Tim Blackmore.

    Includes bibliographical references and index.
    ISBN 0-8020-8791-4

    1. War – Forecasting.   2. Military art and science – Technological
    innovations.   I. Title.

    U21.2.B5 2005   355.02′01′12   C2005-904134-X

An earlier version of chapter 4 was published as 'Rotor Hearts: The
Helicopter as Postmodern War's Pacemaker,' in *Public Culture* 15, no. 1
(winter 2003): 90–102. © Duke University Press.

University of Toronto Press acknowledges the financial assistance to its
publishing program of the Canada Council for the Arts and the Ontario
Arts Council.

This book has been published with the help of a grant from the Canadian
Federation for the Humanities and Social Sciences, through the Aid to
Scholarly Publications Programme, using funds provided by the Social
Sciences and Humanities Research Council of Canada.

University of Toronto Press acknowledges the financial support for its
publishing activities of the Government of Canada through the Book
Publishing Industry Development Program (BPIDP).

Every effort has been made to obtain permission for the use of the illus-
trations in this book. Any errors or omissions brought to our attention will
be corrected in subsequent editions.

This book is for my three families that I love so dearly:
Jesse, Gilda, Russell, and the Bob
and
Richard, Quentin,
and
Joe, Gay, Nedsky, and Mary-Anne

One day the day will come when the day will not come.

– Paul Virilio, *Open Sky*

Contents

Acknowledgments

Here are my heroes and tea makers. From moment one, Gilda Blackmore has been the lodestone guiding my thinking about veterans, war writing, the world, and war. I cannot say how much this always has, and always will, mean to me. In different but irreplaceably powerful ways, Jesse Blackmore has forever been my rock, and Russell Blackmore my guard; both have made a peace in my heart so that I can study war. Chris the Bob has been there with the wonder-grill all along. I cannot imagine doing the work I do without the phlegmatic humour and endless encouragement that Ned Hagerman, my travelling companion on the path to total war, has provided over the last two decades. My other advisers are always with me and are consulted daily: Bob Adolph, Sheila Embleton, John Unrau, Susan Warwick, Barrie Wilson, Harriet Wolff – champions all. To the people who took a chance on an unknown and listened to him with seriousness, Cathy Calloway, Paul Daum, and Dale Ritterbusch, I return the favour with gratitude, respect, and friendship. Phil Beidler's shining example showed me how to begin and how to continue, scholastically and otherwise. I wouldn't have made it here without Richard Swinson and Quentin Rae-Grant – gratitude is a ridiculous word when it comes to acknowledging these two superb people. Joe and Gay Haldeman, astounding adoptive relatives, have made the world marvellously possible. I am gladdened by and lucky in the friendship of Paul Headrick and Kaveh Kardan. This book owes its focus, clarity, and determination to one of the finest co-workers I could wish for

– my editor Siobhan McMenemy – who should be running the world, not editing books. If something is unclear, it isn't because she didn't warn me. I am grateful to Curtis Fahey, careful reader and thoughtful copy editor, who smoothed out the rough places. The people who made me welcome at the University of Western Ontario, particularly Carole Farber, Gloria Leckie, Gill Michell, Catherine Ross, and David Spencer – as well as the original fab four partners in crime: Jacquie, Grant, Ian, and Nick – have my deepest thanks. To John Fracasso and Dave Mills, who somehow managed to keep their warm good humour despite the perpetual angst I brought them – thanks dudes! Finally, the students of Killer Culture have taught me more than they know about war, and also peace – I can't wish for better.

WAR X

Introduction

Why X

What did I hope to gain? More bombs are coming. Dig your holes with the hands God gave you.

– Anthony Swofford, *Jarhead*

Behind this book breathes the great atomic hush – a silence to which the person thinking about war must be tuned. It is dishonest to consider war in the twenty-first century without having to first acknowledge nuclear, biological, and chemical (NBC) war. For more than a decade and a half and across four presidential administrations (Eisenhower, Kennedy, Johnson, Nixon) during the American involvement with Vietnam, the United States considered the use of atomic weapons in southeast Asia.[1] False starts on strategic missile shields in the 1980s and renewed plans for the same in the twenty-first century indicate that humans persist in planning for 'the unthinkable' nuclear war, or more palatably, 'limited nuclear war,' or 'low level nuclear war,' as if any of these things exist. This book is about other kinds of war, yet my words float across the spectre of a quiet earth, devoid of most life, left to whatever insects and hardy animals that can survive such a post-war ecology.

My concern is with humans who interact with war technology. It suggests that humans have engaged and are engaging ever more thoroughly in intimate connections with technology of all kinds in order to extend themselves on the battlefield. In military language, the letter 'X' often indicates a technology, yet to be named – it signifies what is called a 'test-bed,' a place for new equipment to be tried, then advanced or retired. Fighter jets and robotic craft are often first given X designations before they are fully developed. Some projects, like Boeing's X-45, now called the Unmanned Combat Aerial Vehicle,[2] get battlefield trials before they are named. This book's title points at three issues: that humans have been using technology, particularly in twentieth- and twenty-first-century industrial warfare, both to save and to extend themselves. When humans connect to a gun that magnifies their vision, allows them to see, sight, and shoot in the dark, over walls and around corners, they have extended their capabilities: What has been the cost of that extension? As well, X designates a generic human creature who can fit into any machinery and represents the standardization of life for war. The final reason for the X is that so much of the technology I discuss is experimental. There are suits of powered armour, robotic tanks, rotorcraft that can outdo all previous helicopters for agility and power, uninhabited vehicles in the air, underwater, on the ground: all of these create an environment for themselves.

I am interested in the ways that humans come in contact with such machines, how the existence of this technology alters our understanding of who we are, particularly what constitutes the body and self (which I believe cannot be easily separated). In telling the partial story, as all stories are, of recent technology, I also felt that it was right to begin and end with the human story: What is it like to face industrial battle, and to try to return from it? For all that technology seems to organize its own world and existence, I hesitate to use the word 'life'; it is humans who have made machinery into a way of living. In terms of war technology and creativity, this last century was a staggering one. Against artillery humans dug trenches, and to keep the trenches in place, they put up machine guns and then made tanks to confront the machine gun. In the Second World War we put fast low-flying aircraft and personal rockets against tanks: the roundabout continues. If technology is not in control, it is hard to believe that we are either. This book explores some of the missing parts, the zones between things, between the anxiety of having, and not having, technology on the battlefield.

In writing this book I have involved the reader not only in claims made by weapon makers but also in those made in the name of what the chief defence theorists of the United States call the Revolution in Military Affairs (RMA). Once military discussion begins, military acronyms have a way of shouldering into the talk. I'm asking the reader to face some of the jargon because this is the materiel world. Acronyms are a private language of murder and involve the death of people we know, who live around us, or across the world from us. We are all subject to it. I want it to be possible for the reader to get up from the writing, put an acronym into a web search engine, and see what is there, what it is we are using to kill each other.

There have been a number of different kinds of revolutions in military thought in armed forces over time. When I talk about the RMA I'm referring to a recent change in the American military, one that grew from the realizations of Soviet General Staff member Marshal N.V. Ogarkov, who proposed in 1982 that long-range precision strike weapons like cruise missiles, when connected to a military Internet that could gather intelligence from across and above the battlefield, were the new best way to wage war. Ogarkov's understanding went beyond the Warsaw Pact – NATO organization of power that had governed so much of the Cold War. But, if two superpowers weren't locked into a potential European war, what would contemporary and future armies look like? The discussions that ensued in the Soviet bloc and the U.S. military resulted in what is now known as the RMA. It is important to keep in mind that RMA propositions are theories, many would say fantasies, about the way war is now, or in the future should be, pursued by the United States and its allies. For a moment let's float over that new war zone.

It is expected that war will be fought not in lines but in a four-dimensional battle arena. Battlespace, as it is called, is intended to be deep, high, wide, and simultaneous: there is no longer a front or rear (see fig. 1). Computer networks that will talk to each other over a military Internet called the Global Information Grid (GIG) that, with an infusion of $800 million USD in late 2004, promises 'unlimited bandwidth' to its users (Tiboni). The more data flow, the more transparent battlespace is supposed to become, guaranteeing friendly forces informational and tactical dominance of the field. Even before

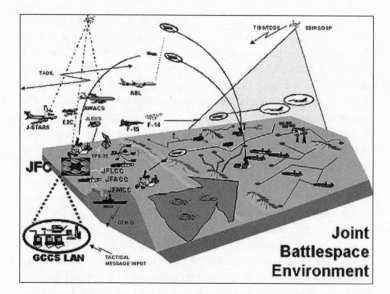

1   **Battlespace**: All war all the time.

the ground war starts, attacks on the enemy's information-gathering and -dispensing equipment begin so as to render blind, deaf, and dumb all radars, radios, telephone exchanges. RMA doctrine manuals and slide presentations sometimes show battlespace as a clear dome under which friendly forces hold full information control. At the top, satellites feed information to airborne flying information centres that act as information hubs for forces at sea (at the left), on the ground (in the centre), and at low or high altitudes (helicopters or F-15 fighters). In battlespace everything will be connected to the GIG so that the forces operate as a net in a linked command structure, not along lines separated by antique interservice rivalries. RMA believers no longer talk about combat but about warfighting, a term used to distinguish the various kinds of activity in which the forces may be involved. While peacekeeping and humanitarian missions sheltered under the title of Operations Other Than War (OOTW) may involve a good deal of killing, that isn't their aim. The goal of warfighting, however, is killing.

These concepts shape the space in which I have written this book. There is no return from a war fought under nuclear, biological, chemical conditions. But the prevalence of those weapons has not prevented a new war world from coming into being. The Revolution in Military Affairs – all lightness, speed, information gathering, information technology, and shared materiel – seems like a cleaner, better way to fight. Yet, in the crush of data about machinery, I have worked not to lose sight of the body. When war persists, the light technology and precision weapons – munitions so smart they're 'brilliant,' according to their makers – give way to heavy power and indiscriminate slaughter. As the U.S. forces prepared for a second ground fight in the Iraqi city of Fallujah in late 2004, talk of precise bombing and surgical strikes became reports of 'pounding,' 'repeated strikes,' 'airstrikes,' 'thunderous [artillery] barrages,' 'blitzes,' and an unofficial comment from an unnamed military commander about '12 hours of overnight strikes' by every available heavy weapon on the ground or in the air (Vick). Added to the butchery is the reintroduction by American forces of napalm to battlespace. Typically a mixture of benzene, gasoline, and polystyrene, napalm is a gluey substance that, once ignited by a charge in the bomb, sticks to flesh, suffocates, and burns its victims to death. Banned by the 1980 Geneva Protocol III of the Convention on Certain Conventional Weapons – an agreement signed by all countries except the United States – napalm use by American forces in Fallujah was reported. The Department of Defense has attempted to refute such claims, arguing that its Mark-77 300 kg incendiary bombs consist largely of kerosene-based jet fuel that, while it sticks to and burns people, does less harm to the environment than napalm (war watchers knew that napalm was back when reports of 'melted' bodies, the signature of chemical syrup burning at hundreds if not thousands of degrees, became public).[3] The linguistic shift from what George W. Bush categorized as the 'catastrophic success' of the initial twenty-one-day march on Baghdad to more conventional images of beating a city and its people to ruins – Winston Churchill called it 'mak[ing] the rubble bounce' – indicate how much the world of the RMA is one of science fantasy.

Because science and war sweat ever more closely together, it is no surprise that science fiction has anticipated war technology. One of the large subgenres in science fiction is military SF, which explores alternate past and potential future wars. Significant thought experiments about the future of warfare have come from, among others, Robert Heinlein's *Starship Troopers* (1959), Joe Haldeman's *The Forever War* (1975) and *Forever Peace* (1997), and Orson Scott Card's *Ender's Game* (1985). Each text's politics is quite different but the imaginings are useful to the military, which has in the past made both Heinlein and Haldeman's books required reading at their top military academies (West Point, the U.S. Air Force Academy in Colorado). There are many powerful antiwar SF novels and a host of SF war romances. It is the latter that have proved so lingering and so deadly.

The RMA and the technological vision it extends to humans as war makers, users, suppliers, and victims is a concoction of fantasy land warfare in which all weapons work as promised in their promotional brochures. When humans are confronted by battle, it makes a certain sense that they would want to have attached to them a variety powerful weapons full of magic bullets, to be able to call on forces from the air and sea that are so astonishingly powerful they kill as many of the enemy as possible (while not harming one's friends), without causing internationally embarrassing amounts of blood to flow, and by example so terrify all survivors that they throw down their arms and fall on their knees in recognition of their wrongdoing. What one might forget in holding any piece of sufficiently advanced technology, however, is the apparently endless hose dragging behind it.

The computerized auto-targeting rifle in my hand, a weapon full of bright munitions that program themselves to explode at the right distance or time, is not as portable as first it seems. The gun's empty magazine requires special ammunition from particular manufacturers countries away: it is part of a supply chain thousands of kilometres long. When I point the gun and look through a thermal imaging sight, take a snapshot using the scope, and e-mail my squad leader a picture of what the gun sees, I connect to a network of information machines hovering around me, supported by vast quantities of electric and human energy. My gun is tied not only to an arms manufacturer and munitions supplier, to the Army supply and logistics apparatus that keeps it operating in case I drop it or it malfunctions, but also to people who make information systems, portable sighting devices, and infrared and thermal vision technology, to those who write software for the chips in the bullets and scopes, and to all the factories that produce those items. What I hold is not so much a single weapon as a nozzle on a fire hose of killing technology. My gun necessarily connects to a war hydrant and, by default, to a whole country's war system maintained at a certain pressure.

Much of the equipment I discuss in this book emphasizes speed, lightness, and portability yet is handcuffed to its designers, manufacturers, distributors, and service departments more than any previous war technology – ever. It is presented by its developers and proponents as a mechanical fantasy come true – weapons that see, munitions that think, rolling and swimming robots that act autonomously, fly for weeks, sometimes months, unattended. These alluring, mesmerizing, promising weapons connect their users to an often invisible pipeline of global defence industries. The machines offer so many slick features that after a while I may have forgotten to ask how much it all costs (not just in dollars but in the quality of all our lives), and what has been surrendered so that they can be made. Instead, I am fascinated by what is next, the new version, the upcoming release. I am in a panic to upgrade. Dropping the weapon means letting go of the whole chain of procurement and supply, design and acquisition, a way of life. A lot of people, families upon families, have and will continue to have jobs because of the weapon I hold. Those jobs let them participate in the fantasy of a better, safer tomorrow in which bad people alone are killed and the innocent spared. They are allowed in on the magic.

In mid-2004 I was interviewed by a television station about war technology. On my way out of the studio, I was told by the next guest who had been watching my interview how wrong I was about the world, how misled. I have this to say to him: I certainly believe in things others don't, but I don't believe in magic. I have written this book because the human body is subject to the laws of physics, chemistry, and biology; the weapons I discuss are part of that mundane world and the real effects it has on all of us. I hope we can remember to remember people.

Chapter One

Crawling Flesh:
The Infant Comes to Battle

The mass shudders; because you cannot suppress the flesh.
– Ardant du Picq, *Battle Studies* (1868)

This is a story about soft flesh and the meeting it has with industrial war technology. It is a slash, an open wound – we're still working on the flesh story, trying to stitch it closed. When flesh is spread apart by humming shrapnel, ball bearings, cones of metal, or is consumed by unquenchable white phosphorous, the soldier sees that industry is the winner. The flesh story is inscribed on countless millions of bodies, most gone by now, eaten by the earth or conjured away by high-explosive alchemical technology. The mass of bodies lying nameless in battlefield graveyards is remembered in words that pass from soldier mouth to soldier mouth on a trail of saliva leading to the softest flesh in our bodies: brains. The brain seems invincibly armoured by its remarkable bone cask, but we'll see that the brain can become the centrepiece of damage. After watching bodies shredded by industrial warfare, soldiers understand their own bodies differently: the appetite for survival brings with it other cravings. The military uses an enormous stamping mill to press soft bodies into hard-core corps, but even that cannot armour the self against war. Hardening is always the way it begins, though.

Civilian bodies approaching war must be prepared for the hardships of combat. Would-be soldiers imagine good fortune, gifts of respect and privilege the culture will bestow upon their bodies, small metal markers to be worn on the chest, signifiers of other pieces of metal they faced and perhaps absorbed during battle. Volunteers and draftees alike spend much of their time hoping that they can reclaim their bodies once the fight is over. The military understands that fresh bodies are, in themselves, useless. The soft civilian body must be replaced by an infinitely tougher, trained creation, one able to withstand and perform in the war machine. The soldier may arrive in the military without any idea of the narrative that is about to unfold. Paul Fussell became a soldier in the Second World War because 'when I was sixteen, in junior college, I was fat and flabby, with feminine tits and a big behind. For years the thing I'd hated most about school was gym, for there I was obliged to strip and shower communally. Thus I chose to join the R.O.T.C. (infantry, as it happened) because that was a way to get out of gym, which meant you never had to take off your clothes and invite – indeed, compel – ridicule' (*Boy Scout* 254–5). Shame of the body's last fledgling aspects of its youth inadvertently drives Fussell into combat. Ever savagely honest, Fussell the man attacks Fussell the boy, not only for being fat and feminine but for doing something as stupid as joining the army weight-loss plan. But that plan is effective – the army's first task is to discipline and toughen the body, preparing it to serve in and survive almost all weathers and conditions, including the bone-cracking winters of Europe and North Korea or the hothouses of the Pacific islands and Vietnam.

Like Fussell, Second World War German Wehrmacht veteran Guy Sajer found his body treated brutally as it was trained for war: 'Without a moment's hesitation, we were all stretched out on the sandy soil. Then Captain Fink stepped forward and, like someone strolling down a beach, walked across the human ground, continuing his speech as his boots, loaded with at least two hundred pounds, trampled the paralyzed bodies of our section. His heels calmly crushed down on a back, a hip, a head, or a hand – but no one moved' (162). Sajer has already been at the Russian front, his body milled by the youth brigades and regular German army, but in order to join the Wehrmacht he must prostrate himself further. Three men will be killed by Wehrmacht officers during Sajer's training in order that their bodies not weaken the whole. The captain's narrative is about force and pressure. His pace is deliberate and inevitable, designed to produce fear and the desire to avoid a boot heel, a measure of the bullets and shells to come. Fink's boots teach the soldiers' bodies about control, about signing individual reflexes over to the collective war body. Hardening makes the soldier feel superior and invulnerable: now there is muscle armour where there was soft flesh before. Fussell reflects on the lessons his body learned in 1944: 'If in the First World War you're one of over four million uniformed Americans, you're actually pretty anonymous. But if in the Second World War you're one of sixteen million, you're really nothing' (*Wartime* 70). The insignificant self disappears into a mass. Illusions about the body as personal property are flattened by the press of the captain's boots over the flesh that exists only to be landscaped by the military.

The necessarily brutal quality of the training that divorces the soldier from the self prepares the new body for a new kind of environment in which there is little (if any) sleep, food, water, and medical care. It is a world that requires a body able to perform repetitive actions (field stripping – the process of cleaning or fixing a weapon while under fire – loading and reloading weapons, firing automatically, servicing machine-guns, mortars, and artillery), in what military historian John Keegan characterizes as 'an undiscovered continent, where one layer of the air on which they depended for life was charged with lethal metallic particles,' a barrage of projectiles 'whose presence rendered battlefields uninhabitable (giving them that eerily empty look which, to an experienced twentieth-century soldier, is a prime indicator that danger lies all about)' (306). The air now growls with swarms of black-hot metal bees flying on industrial high-explosive wings, impressing upon the body the urgency of stillness: '*Stillgestanden*,' was the first command Sajer learned.

When, writing about the undeclared Korean War, American novelist Stephen Becker says, 'All war is now total war,' he suggests two senses in which this is true (133). The first is that wars' participants can no longer easily be divided between combatants and non-combatants. Typically, 'women and children,' perceived to be innocent unless they carry weapons, are spared both on and off the battlefield. But after the death camps, German atrocities in Russia, the indis-

criminate bombing of London during the Blitz, the Allied immola-
tion of Hamburg, Dresden, and hundreds of Japanese cities, the idea
of non-military – and so illegitimate – targets disappeared. The second sense in which 'all war is now total war' is that there is no front
line, no rear, no place to which it is safe to retreat. Escape is impossible since everything has become battlespace.[4]

Twenty-first-century warfare has surpassed the technological battlefields of the Second World War and Vietnam and replaced them
with a space in which digital warfare, robots, electronic weapons,
and the Internet combine to form a new way of fighting that is
entirely dependent on factors itemized as Command, Control, Communication, Computing, Intelligence, Surveillance, and Reconnaissance (C4ISR). According to contemporary land-warfare specialist
Michael Evans, the sphere in which combat occurs is an area
'quickly fractionalized by firepower. Unpredictable movement and
confusion often dissolve unit formation. Combat resembles an eddy-
ing tide, not a single incoming wave' (41). The wave looks back to
the Second World War, an era where friendly and enemy lines were
reasonably delineated; even the so-called 'lightning war,' or *blitzkrieg*, looks stable by comparison with battlespace whirlpools.
Twenty-first-century soldiers might overcome the panic of existing in
a fluid war zone by focusing on personal agency: the soldiers' personal weapon, personal body armour, personal skin beneath that
armour. These will all turn out to be illusory, but fear makes any reassurance attractive.

The skin is a bag for the body, also the soldier's first line of defence
– but a weak one, easily breached. The new soldier doesn't know yet
how easy it is for industrial warfare to penetrate the body, to erase the
separation between the person and the environment. For the green
trooper, the war machine may possess the soldier but the skin defines
the edge of that possession. As time goes on, the soldier becomes fix-
ated on skins, envelopes, bags, barriers, because they keep the soldier
in and together and the world external and apart. There are a variety of
bodies on the battlefield: there's the war body that has been trained to
work in concert with other bodies; the war body that has been
destroyed; and the war body that is subject to unbearable physical
conditions. In the first and second instances, the soldier either has

company or is insensible of it, but in the third body, the soldier is both profoundly alone and aware. The sufferings are particular to the individual and cannot be shared among the group. Confronted by the Russian winter, the German army froze into tiny units: with temperatures dropping below minus 30° Fahrenheit, the soldiers had to survive first, fight second. New arrivals to the front saw only piles of cloth in foxholes and 'needed to be accustomed to this strange mode of existence to know that beneath these mounds of cloth subtle human mechanisms were managing to survive and garner their strength' (Sajer 75). The extremes of temperature make it remarkable that any should survive, let alone be able to consider fighting. Sajer's odd phrasing about machine people suggests the soldiers are complex, perhaps autonomous, devices. The complex original is lost under a messy pile of fabric, second skins. The soldiers are alone in their holes, barely functioning, inching along the edge of existence.

Such crawling humans are familiar to historian John Keegan, who reflects that the fighting in the 1916 Somme campaign was 'as earthbound, snailpaced and softskinned a business as it had been for the two hundred preceding years' (285). Gas, machine-guns, high explosives, and advanced artillery drive men onto, then into, the mud. Thin human skins exposed to industrial weapons can survive only if cloaked by trenches, thick mantles of earth. The helmet, the most rudimentary piece of armour, protects the soldier, but it also interferes with being human, as Wehrmacht soldier Guy Sajer recalls: 'It is difficult even to try to remember moments during which nothing is considered, foreseen, or understood, when there is nothing under a steel helmet but an astonishingly empty head ... There is nothing but the rhythm of explosions ... and the cries of madmen to be classified later, according to the outcome of the battle, as the cries of heroes or of murderers' (184–5). Sajer's steel helmet figures repeatedly in his autobiography, always cutting off his thought, imprisoning him in a miasma of fear. His recollection of a time spent in body armour is blunted by it – he isn't able to think about the war world in which the present and future are abolished. He is one of Keegan's snails, slowly trailing along, terrified and empty. The snail's shell prevents not only memory and understanding but also language and hearing. The barrage is the force that governs the battle. Personal actions cannot be judged: after it is over, the winning side will determine who are the saviours, who the murderers. While the battle is under way, all are equally insane.

Sajer's account of battlefield chaos isn't particular to Germans fighting on the eastern front, Italians in the First World War, Americans on Pacific islands in the Second World War: all industrial battlefields of the twentieth century demonstrated to the soldier the body's spectacular vulnerability to new weapons. Many soldiers will have been raised on tales of battlefield heroism, accounts that nourish the sense that humans control wars, that one determined human can change the course of events for a mass of others. According to Second World War veteran Paul Fussell, belief in heroic agency breaks down gradually, if predictably, through

two stages of rationalization and one of accurate perception:
1. It [violent death] can't happen to me. I am too clever/agile/well-trained ...
2. It can happen to me, and I'd better be more careful. I can avoid the danger by watching more prudently ...
3. It is going to happen to me, and only my not being here is going to prevent it. (*Wartime* 282)

The first steps suggest that the soldier is still in charge. Bad things unfortunately happen to other people, but they won't happen to the individual who is more worthy or better adapted. When the fearful soldier is forced to admit that good people who know what they're doing in battle *do* get hurt or killed, then the resolve must be to do better and to tread more carefully. In the end these two steps are fraudulent thinking. The *only* way to prevent being hurt is to get away, to set aside cultural rewards (the title of hero, physical medals) and even cultural membership (forfeiting the post-war right of being a civilian) and desert. But such a disappearance is difficult to effect. If one's head is, as Sajer has posited, completely vacant during battle, then planning an escape is infinitely remote. How will the terror-stricken human manage to elude the state, a task difficult even at the best of times? Somehow the soldier must engineer an acceptable form of survival. He or she must be hurt (but not killed) sufficiently badly to be returned home, honour intact. In the First World War, British soldier Robert Graves made just such a deduction and arrived at the following formula:

> I went on patrol fairly often, finding the only thing respected in young officers was personal courage. Besides, I had cannily worked it out like this. My best way of lasting through to the end of the war would be to get wounded. The best time to get wounded would be at night and in the open, with rifle fire more or less unaimed and my whole body exposed. Best, also, to get wounded when there was no rush on the dressing-station services, and while the back areas were not being heavily shelled. Best to get wounded, therefore, on a night patrol in a quiet sector. One could usually manage to crawl into a shell hole until help arrived. (*Good-bye* 111)

Graves has moved through Fussell's three points. He has decided how best to thread the chain of logic, chance, and disaster. His linguistic repetitions are a map of his plans: one reason drops into place after another. Rather than aiming to kill on ineffective night raids, he points his body at the enemy, aiming to be a successful target without being killed. In his culture's eyes, Graves's heroism will make him a winner: this is the valued quality, whether or not it has any battlefield merit. In his construction of events, he will be treated, saved, and, if the wound is bad enough, sent home. Further, he will be done with the war because *he* took action, not because of some unplanned battlefield calamity. The spectre of control denies the reality of war's chaos: agency is a wisp of smoke the industrial battlefield blows away, and the body often goes with it.

In order to keep soldiers on the battlefield, they must be trained to believe that they can successfully take charge in combat. The rifle put into each soldier's hand offers the individual great reassurance. And, as the rifle extends the arm's reach, so the bayonet extends the rifle's. Yet, according to military historian Joanna Bourke, the bayonet is a next to useless combat tool (90). Bayonet duels are extremely rare, and killing with edged weapons, even in the Napoleonic era, even rarer (Keegan 264). If these facts are so, why is bayonet practice so prized by armies? In part, it is because there is a belief that bayonet training generates a killing spirit, but more it is because the bayonet personalizes the rifle and the battlefield (Grossman 126). The reality is that the most effective killers in industrial war are high-explosive artillery shells, mortars, and bombs dropped from aircraft. But these weapons are out of most infantry's hands (with the exception of mortar crews – often the first to be fired on by other infantry).

High explosive has reorganized bodies more forcefully and in greater quantity than any other materiel. A U.S. Marine fighting on the Pacific islands, Eugene Sledge, had a lot to loathe, but he put at the top 'prolonged shelling [that] simply magnified all the terrible physical and emotional effects of one shell. To me, artillery was an invention of hell. The onrushing whistle and scream of the big steel package of destruction was the pinnacle of violent fury and the embodiment of pent-up evil. It was the essence of violence ... I developed a passionate hatred for shells. To be killed by a bullet seemed so clean and surgical. But shells would not only tear and rip the body, they tortured one's mind almost beyond the brink of sanity' (74). Sledge's distinction between the bullet fired by one soldier with one rifle and the shell thrown by an enormous gun miles away is intriguingly medical. The bullet, like a scalpel, is surgical, precise, personal: one operation at a time. But the shell commits arbitrary collective dismemberments. The bullet marks itself by small puffs of dust, or even alarming zipping noises, but the shell is too large to comprehend. Sledge's terror, his anticipation of trauma so bad that the body cannot in any way withstand it – even the worst bullet wound will leave some of the body behind – makes the shell the agent of bodily transgression and shame. At the same time, the infantry embraces the shell. Historian John Keegan notes that, by the end of the First World War, soldiers 'had learnt, and were glad, to walk as close as twenty-five yards in the rear of a boiling, roaring cloud of explosive and dust, accepting that it was safer to court death from the barrage than to hang back and perhaps be killed by a German whom the shells had spared' (248). As artillery gunners improved, they learned to 'walk' the barrage forward, slowly moving the line of fire towards the enemy trenches. It is this shield of high explosive that the soldiers are pleased to stand behind. The very quality the infantry hates in the shell, the wholesale pulverizing of flesh, makes them equally grateful for it. The soldier can accept the most dreadful weapons if, when they kill the enemy in sufficient numbers, the body can be spared.

Part of the terror of being bombarded is the shell's arrival from an unknown, remote location. The standard Second World War American heavy artillery field piece (the M115 howitzer) could throw an 80 kg high-explosive shell sixteen kilometres: they literally dropped out of nowhere and could wholly eliminate the body. Soldiers observing artillery victims report the surreal nature of seeing a person be transformed into a red mist, or simply vanish. One Vietnam veteran recalls a grunt who got up one second early: 'As he stood up one of the artillery rounds hit on the other side of the road from him and it just atomized this guy. We couldn't even find a fabric of cloth. He was just gone' (quoted in Lehrack 165). The disappearance seems clean: there is no screaming human left behind, no half-corpse, no travesty of a body to cope with. But the complete emptiness, the vanishing act that heavy weapons perpetrate on the body, is even more terrifying. Where there was a fellow being, now there is nothing – or just, as Fussell suggests, 'spots' (*Wartime* 278). The Vietnam veteran's language is also shattered by the blast: the 'fabric of cloth' is a linguistic shudder which, more than twenty years after the event, reflects the explosion. Even in quiet recall, the veteran cannot summon connected words. Removing the body is shocking, fragmenting it is equally bad. Artillery damage creates broken words: language doesn't stay coherent when shells explosions turn soldiers' bodies into projectiles, and soldiers are 'hurt, sometimes killed, by being struck by parts of their friends' bodies violently detached' (Fussell *Wartime* 270). Here again is language turned sideways: an abrupt wrench annihilates wholeness, the integrity of the body, the idea of safety. To be wounded by a shell is bad, but to be hit by deadly airborne bone fragments of a fellow soldier is even more grotesque. The soldier has begun to learn how easily the skin can be invaded, but he still must learn that death is not always a visible penetration.

High explosives are known for their ability to kill even those closeted in deep bunkers. A shell exploding near a bunker entrance creates a sudden vacuum, an event observed during the First World War. Known as 'blast over-pressure,' the effect kills without producing messy corpses: over-pressure creates, according to John Keegan, 'vacuums in the body's organs, rupturing the lungs and producing haemorrhages in the brain and spinal cord' (264). Bunkers full of apparently healthy bodies, men sitting upright, paused in tasks, or perhaps napping, turn out to be dead – bleeding from the eyes and ears. In the centre of winter, fighting the Wehrmacht in Alsace, Paul Fussell stumbles on an entire German squad that 'looked like plaster simulacra excavated from some chill Herculaneum,' which has been killed in place by an artillery bombardment (*Doing Battle* 134). The dead men, rimed with ice, are frozen in position: it's a tableau that amazes even Fussell, the self-professed sceptic. Artillery lessons are fearful ones: the warnings (whistles, screams) of an incoming shell are signs of panic; there is no safety above ground; going underground may avert dismemberment but cause even deeper bodily transgressions. The thin, soft-skin bag is torn open by an arbitrary but malign hard force that illuminates the body's fragility. The big guns focus the soldier on the permeable body. One Second World War German soldier on the eastern front can no longer eat plums because they are too reminiscent of other horrors: 'Burnt human flesh there before you ... you throw the plum away, for in your hot fingers it has become too soft, and you are for a thought-provoking moment too soft yourself' (Fritz 144). Revelation and connection to softness, the ease of damage, cools the veteran's appetite. Human survival guaranteed by food has been jeopardized; the association of the body with the plum is horrid enough that the soldier must rid himself of that item so rare and marvellous at the front – fresh fruit. Fewer things are as hard to come by. Either way the plum is about death: starving is always nearby; becoming soft, becoming food, is disaster.

Sledge, moved from one dreadful Pacific island combat post to another, also reflects on the soft body. Many of the islands are only a screen of dirt over coral and rock. There is no fresh water, no way to leave foxholes that have been blasted into the coral, no way to bury the dead, dispose of rotten food or human wastes. The soldier sits between fights and watches war debris cooking in the tropical heat until the battlefield becomes a vast sewer, a truth Sledge must face: 'If a Marine slipped and slid down the back slope of the muddy ridge, he was apt to reach the bottom vomiting. I saw more than one man lose his footing and slip and slide all the way to the bottom only to stand up horror-stricken as he watched in disbelief while fat maggots tumbled out of his muddy dungaree pockets, cartridge belt ... We didn't talk about such things. They were too horrible and obscene even for hardened veterans' (260). Softness is one problem, but rottenness is something else. The soldier eating a plum is dismayed, but here the victim is unable to function once immersed in a pit of liquid flesh. Soldiers, veterans and newcomers alike, haul each other out of these sump holes, shaking, speechless, beyond language. There's an inevitable quality to the slide that reinforces the horror. Nothing can stop the body once it has begun to slip except for the mass of flesh and bone at the bottom of the hole. The live soldier must instantly be distinguished not just from the dead he wallows in but from the all-too-live maggots that appear everywhere. It seems as if the soldier will be consumed alive by the maggots if he doesn't climb out quickly enough.

Soft bodies, ripe and rotten bodies, plague Second World War German soldier Guy Sajer, who cannot contemplate flesh: 'Strange bodies made me uneasy, almost sick. As soon as I saw naked flesh, I braced myself for a torrent of entrails, remembering countless wartime scenes, with smoking, stinking corpses pouring out their vitals' (355). The skin is only a shiveringly fine barrier. It is no longer any guarantee of safety, no longer a way to preserve the body's cohesion, no longer a promise of solidity. It has almost ceased to be a surface – clothing has taken over that role. Skin might as well not exist. The dying body isn't passive, but a kind of flesh cannon that will bombard the sickened viewer with corrupt entrails: the body has become a gas-filled explosive shell that, when detonated, drives its victims insane. Sajer would understand Sledge's near insanity when he is forced to dig a mortar emplacement through an all-too familiar impediment: 'The next stroke of the spade unearthed buttons and scraps of cloth from a Japanese army jacket buried in the mud – and another mass of maggots. I kept on doggedly. With the next thrust, metal hit the breastbone of a rotting Japanese corpse ... The shovel skidded into the rotting abdomen with a squishing sound' (277–8). A veteran of three Pacific landings, including one of the most wretched campaigns in a war zone infamous for wretched campaigns, Sledge cannot summon his hatred of the Japanese to make this task more bearable, nor can he continue it.

There can be no reconciling with the body's contents: the soldier has been profoundly betrayed. What does the dead body have, apart from a little breathlessness, that the live one doesn't? How are they different? Loathing and shame overcome the owner of what might in a moment become an unbearably messy corpse. How will the soldier prevent his own death from being so public, obscene, and open to scrutiny as the ones in the Pacific campaigns? Battlefield industry does everything to breach a body now terrified of itself. Sledge's Second World War disgust with flies that wander from corpses to his canned military C-rations (144) is transformed into a Vietnam veteran's rage:

> to this day ... I go out of my way ... I go extremely out of my way to kill a fly. I don't know what that fly is on when I go for it. It doesn't make any difference. I'll get a fly with a chair, a shoe, a book, a flyswatter, my hand. They just get to me. I'll tear a screen out to kill a fly. And I feel so ... so terrible after going through a routine like that I can't stop it. And sometimes I don't even realize that I've done it until after it is done. I'll find a screen bent or something broken, a fly on my windshield, I'll just give it my personal best ... I think that I would wreck a car and sacrifice safety to kill a fly. (Quoted in Lehrack 84)

The veteran's near-incoherence recalls the 'fabric of cloth,' that moment when the body completely disappears in a shell explosion. The fly survives, perhaps even causes serious injury, and claims the body: at all costs that possession must be stopped. The fly, the maggot's kin, is a representative of heat, battlefield death, bodily transgression (the body is left unburned or unburied), and loss of control. The veteran can't control the obsession that has taken over: the fly is in charge. Cultural critic and philosopher Elaine Scarry argues that 'what is remembered in the body is well remembered ... those new alterations are carried forward into peace' (112–13). The body remembers its shaming and acts automatically. It remembers how artillery shells, the tropical sun and the Siberian winter, high explosives, napalm, and land mines treat it. It remembers its public evisceration, being laid open for inadvertent anatomy lessons that teach the living how much is stuffed inside them.

What becomes abundantly clear to a soldier at the forward edge of battle is that the culture has already written off his body: he is what the British Army during the First World War, speaking of its seven thousand daily casualties, would call 'wastage.' A laconic Second World War British tanker in North Africa sums up the soldier's position: 'You go in [to the line], you come out, you go in again and you keep doing it until they break you or you are dead' (quoted in Ellis 239). There is no way out of combat unless the soldier can engineer the kind of luck that Robert Graves hoped for in the First World War. Graves knows the tank veteran's plight: one can only go forward into the war and presumably through it to a mythical realm of peace. Unless there is a miracle, the soldier likely moves towards death. Allied soldiers who had finished the eleven-month post-D-Day push across Europe from June 1944 to May 1945 were then regrouped for the final assault on the home islands of Japan. The U.S. Marines fighting in the Pacific and the Allied veterans of the European campaign had few illusions about the odds of their survival in such an assault (it was thought that all of the first ten amphibious waves of soldiers attacking the island would be killed). Contemplating the invasion, U.S. Marine and later military historian William Manchester wrote famously, 'Thank God for the atomic bomb,' a sentiment echoed by millions of other soldiers (250). The atom bomb was the kind of miracle a foot soldier could only dream of: two shots, two days, and the war was over. It seemed incredible and for the men spared the assault, wonderful. The ethics of the situation would have to wait. The atom bomb didn't erase the need for obedient soldier bodies, and arguably increased it given that post–Second World War soldiers confronted the prospect of atomic war. Keeping soldiers in the line continued to present a problem.

The military drills into the soldier the shame of quitting and fosters deep ties between soldiers in a squad. During the Second World War, the army learned that soldiers don't generally fight for abstract concepts like honour or democracy: they fight for each other, to protect each other, and to demonstrate to themselves and the group that they are worthwhile people. Soldiers who agree to these terms go into combat with death in mind – they accept that their bodies may be required in order to prevent shaming themselves or their groups. In

Korea, a Marine officer requests that a soldier kneel and act as a platform for a light machine gun: 'I had used his living body as a kind of sandbag, and he had been killed while performing this strange role' (Russ 289). The officer laments that, while the man's death 'was extremely painful to me,' he concludes that using a human gun mount was not wrong. All the soldiers accept these conditions – at least that's one story. There is an ongoing struggle between the soldier's panic at lack of personal agency, at the fragile skin preventing the body from destruction, and the soldier's wish not to be seen as unworthy in the eyes of others in the squad. Fear of bodily shame by violent death slams up against the prospect of shame in front of the community. Between these two is the human paralysed, kept in place by the military that disposes of bodies when necessary because it understands them and deals with them in the aggregate. The personal story of the soldier gun mount is transformed into a battlefield narrative complete with mass slayings. Having survived night combat in Korea, a Marine wakes up to see that 'there were bodies everywhere. None of us really understood what had happened until it got light. Then we saw there were so many corpses they actually changed the contour of the terrain' (Russ 127). The intimate stories of the plum, the Marine in a Pacific island death trap, when multiplied and added together, become a landscape of death. The soldiers' bodies are the territory. Their deaths have not guaranteed them a victory, nor has the fact that their bodies have remade the land made a difference to their cause. The Chinese and American militaries know that bodies are only gun mounts, but the soldiers have to wake up to see the corpse world around them to understand how their personal terrain has altered.

The miracle of the soldier is this very willingness to surrender the body. The culture teaches its children that exchanging the body for freedom in its many significations is legitimate and, more, desirable. Paradoxically, the body is surrendered to the state that has guaranteed its safety. The civilian who is about to become a soldier may join the army for many reasons: because of a military heritage, out of pride in service past generations have given, because of a pervasive belief that the military is a positive force in life, to get expensive schooling, for money. In these ways, the non-combatant is already mobilized, having formed part of a standing army even while living a civilian life. When the killing begins and the fractured bodies and sealed caskets start to come home, the army begins to run short of volunteers. Then the draft begins, bringing with it various forms of civilian resistance. A short list of rebellions indicates that not all is well even in wars of supposed just causes: during the American Civil War there were four days of severe draft riots in New York in 1863; in the French, Russian, and then German armies in 1917 and 1918, mutiny; in Vietnam (perceived to be a popular cause for at least half of the war), the rise of 'fragging' (originally the use of fragmentation grenades, but ultimately any weapon, to kill over-eager officers who endanger the troops as they attempt to survive a lost cause) and, late in the war, local troop and sailor mutinies, civilian draft-card burnings, emigration to other countries. If they wish to opt out, civilians face the state's enforcement apparatus – police, jails, work camps, and, until recently, execution.[5]

Soldiers-to-be hear selected narratives that come, most tellingly, from survivors, from people who like to remember particular stories (and perhaps forget others), from people who admire heroes, from heroes themselves, and from heroes represented by the culture's artifacts. The dead don't have their say, although the living create other stories about them, stories that include war monuments, poems about the valorous dead, posthumous medals that please the living. The wounded are put out of sight. Becoming a soldier, even if it requires surrendering the body, is attractive until battlefield realities become apparent – then the individual is in trouble. A lifetime of cultural definitions of heroism slam into the truth of combat and the internal world tilts at the impact. The most important questions about personal fortune (Can I survive? Will I? Why me and not others?) force the soldier to review the way the family, local community, religion, medicine, state have provided an explanation for the way the world operates. On the battlefield the soldier moves in a continuum with outright rebellion at one end and surrender at the other. The odds are for surrender: not only has the army prepared for the shock to the self, but the small unit, the new family, reinforces the values of prewar home. The individual is the site of a fierce struggle because, as Second World War veteran Paul Fussell's three stages indicate (I can't be hurt; I *can* be; I *will* be and must escape), personal survival continues to be important. Perhaps for the first time the soldier examines the underpinnings of life in the context of borderless panic, helplessness, and a sense of being trapped. Panic can become rage that turns against any number of people or forces (women at home, the military brass, politicians, other soldiers). When the rage turns against enemies, atrocities can begin.

Alive and dead the soldier body is a site of contest, as First World War veteran Robert Graves points out: 'As for atrocities against soldiers – where should one draw the line?' (*Good-bye* 153). The confusion of innocent with combatant, the widespread attacks on civilian populations in the twentieth century, has perhaps eclipsed what occurs between soldiers on the battlefield. It may be hard to determine a combat atrocity when the very act of placing a human on an industrial battlefield is inherently an atrocious one. Battlefields are supposed to be governed by sets of rules or laws including the Geneva Convention of 1949 and two following 1977 protocols (to which neither the United States nor United Kingdom have been signatory), as well as the Standing Rules of Engagement (for Canadian Forces, simply Rules of Engagement), which are enforced by military courts following the Uniform Code of Military Justice. These are orders that indicate when it is legitimate for soldiers to fire and how to define and treat prisoners of war. Against these rules presses the sensory overload of combat. Second World War veteran Guy Sajer suggests that 'the almost drunken exhilaration which follows fear induces the most innocent youths on whatever side to commit inconceivable atrocities' (234). Having survived yet another assault, barrage, immersion in the pit, the body experiences an insane joy so strong it's as if the soldier leaves the skin and observes events from outside. The Geneva Convention states, for example, that prisoners who have surrendered will not be summarily executed. First, however, they have to surrender successfully.

Paul Fussell recalls a familiar 'turkey shoot,' in which German soldiers trying to surrender from the depths of a large bomb crater meet ecstatic American youth: 'Laughing and howling, hoo-ha-ing and cowboy and good-old-boy yelling, our men exultantly shot into the crater until every single man down there was dead. If a body

twitched or moved at all, it was shot again. The result was deep satisfaction ... If it made you sick, you were not supposed to indicate'
(*Doing Battle* 124). Since the infantry soldier cannot control attacks
by artillery or bombers, there is a craze for command over the body.
The bomb crater was created by a uncontrollable, arbitrary force, so
it is only fitting that it become a site of torment. The torturer appro-
priates the victim's body, takes it over until it has been drained of
life, gets pleasure from exacting penalties on the enemy's body. The
bomb caused terror: now it is the soldiers' turn. Fussell's recounts
that later the story is commuted to an 'amusing narrative,' where rep-
etition and group agreement about it erase personal guilt. In Fussell's and Sajer's accounts, beating enemy bodies down, transforming them through brute repetitious hammering, obscenely shaming
them, allows for some strange reconstruction of the survivors' bodies. Fussell's unblinking record of what he saw is matched by his
famous essay about Marines taking Japanese skulls and body parts in
the Pacific campaign ('Postscript [1987] on Japanese Skulls' *Atom
Bomb*): he refuses to be shamed into lying by offended Marines who
argue that the Japanese deserved everything they got.

Leaked to the press in the late spring of 2004, pictures from Abu
Ghraib, the American-run prison in Iraq, were ugly not only because
of what they showed but because they suggested that the United
States was knowingly engaged in systemic torture. The picture of a
hooded man standing on a box with wires attached to his penis, toes,
and fingers was notably diagnostic: ' "Was that something that [an
MP] dreamed up by herself? Think again," says Darius Rejali, an
expert on the use of torture by democracies. "That's a standard torture. It's called 'the Vietnam.' But it's not common knowledge. Ordi-
nary American soldiers did this, but someone taught them" ' (Barry et
al.). The struggle is on for possession, or in this case, rejection, of
enemy bodies. While the American Military Police committed the
acts and took pictures of them, they have argued that they acted
under orders from Military Intelligence. Seymour Hersh, who broke
the story simultaneously with the CBS television network, argues
persuasively that the torture was part of a black project known as a
Special Access Program that could operate only with the ongoing
imprimatur of Secretary of Defense Donald Rumsfeld. Shaming,

terrorizing, beating, sometimes killing Iraqi prisoners was seen as a way to gain information. Once the pictures escaped, Military Intelligence, Abu Ghraib's commander Brigadier General Janis Karpinski, the Central Intelligence Agency (CIA), Donald Rumsfeld, and George W. Bush all tried to hand the bodies back, despite Bush's memo of 7 February 2002, which stated: 'I hereby determine ... that none of the provisions of Geneva apply to our conflict with al-Qaida in Afghanistan or elsewhere throughout the world' (Associated Press). Less famous but equally significant is the eerie case of Nagem Sadoon Hatab, an Iraqi prisoner at Camp Whitehorse, located near Nasiriyah. In early June 2003, needing an empty cell at Whitehorse, Major Clarke Paulus specifically ordered Hatab to be hauled outside by his neck – he died within seven hours. When Army pathologist Colonel Kathleen Ingwersen performed the autopsy on Hatab and found a broken neck bone, she confirmed that the dragging had killed him. That bone then disappeared. The bizarre now piled upon the strange: Hatab's larynx reappeared in Germany where his corpse had been examined, but his ribcage was sent to Washington, D.C. In court, the Institute of Pathology was unable to explain exactly what had happened. The military judge handling the Paulus trial barred the pathologist's evidence as suspect; with the pieces of body went the aggravated assault charge against Paulus – it was reduced twice to assault and battery, then to dereliction of duty. In the end, Paulus was discharged from the Marines. The bone may have vanished, the case may have vapourized, the ribcage may have travelled alone to another country, but every bodily removal demonstrates that the military owns the corpse and is master of its use.

One Marine dedicated to recounting where the bodies went is Second World War veteran Eugene Sledge, who remembers having fallen into a kind of trance when looking at, and then beginning to covet, a Japanese corpse. For years he has watched men collect fingers, hands, and gold teeth from Japanese corpses. In a fog he drifts towards the radiance of a dead body: 'I noticed gold teeth glistening brightly between the lips of several of the dead Japanese lying around us. Harvesting gold teeth was one facet of stripping enemy dead I hadn't practiced so far. But stopping beside a corpse with a particularly tempting number of shining crowns, I took out my kabar and bent over to make the extractions' (123). The Ka-Bar, a standard issue knife with an 18 cm blade, was a multi-purpose tool for most battlefield tasks; rarely used for killing, it had typically household jobs – opening cans and boxes, cleaning gear. Here it becomes a dental tool in the way that a pair of pliers takes on a new aspect in the hands of a torturer. While the gold is the bait, it is the idea of controlling another body that attracts Sledge. Gold and bone are two durable reminders of what did and didn't happen: one soldier body survived, the other didn't. The souvenir represents rage at the enemy, the shame of the body, and being alive. Seeing him possessed, medic Doc Caswell gradually dissuades Sledge from taking the teeth. But Doc has to work hard at his task: first he suggests that Sledge's parents would be unhappy if they knew what their son was doing, only to have Sledge demonstrate the rationality in atrocity when he counters, 'My dad's a doctor, and I bet he'd think it was kind of interesting.' Ultimately, it is by appealing to the literal filth of the matter – 'The germs, Sledgehammer. You might get germs from' the teeth – that Caswell persuades Sledge to settle for taking the collar insignia. Doc has had the presence of mind to try two tactics, and the second, for Marines stuck on waterless coral atolls with their own sewage, is convincing. Appealing to what his parents would consider correct is useless. In the haze of atrocity, it is the insane (as if the years in combat don't endanger the soldiers infinitely more than germs will) that makes sense.

Fussell interprets the removal, curing, and photographing of Japanese skulls as evidence that 'the marines were proud of their success in humiliating, punishing, and finally destroying an enemy who, violating a quiet American Sunday, had dared to bomb Pearl Harbor' (*Atom Bomb* 48). The skulls constitute hard evidence that enemy bodies couldn't stand up to the American soldier, and if the enemy thought America was a vulnerable body politic, that was being disproved skull by skull. At least one skull was returned home to a soldier's sweetheart who was duly photographed with it, and the photograph was published in a 1943 *Life Magazine* (see fig. 2).

Twenty-five years later, Michael Herr, living with the troops in Vietnam, was shown countless soldiers' photograph albums of flesh souvenirs, including severed ears, fingers, penises, corpses posed in

2   **Remember It Well**: This photograph published in *Life Magazine*, in 1943, bears the caption reading 'Arizona war worker writes her Navy boyfriend a thank-you note for the Jap skull he sent her.' Proof the body could be overcome.

sexual positions. After a while, he concluded only: 'At least pictures didn't rot' (199). Taking physical trophies, boiling them in lye (as occurred with the bone souvenirs taken on the Pacific islands during the Second World War), or photographing them (more common during the Vietnam War) is proof that the enemy can be shamed. Marines on the Pacific islands 'would come upon prisoners whose hands and limbs the Japanese had severed and then reassembled, "laid out in the shape of a man'" (Linderman 159). Why reassemble a body, why stuff soldiers' penises into their mouths, if not to show the Americans the limitless degradation of the defeated body? The reconstructed body acts as both a warning and a road sign: ahead lies shame beyond death – the enemy will master the body. Where there was a tongue, now there is a useless penis, a double castration and silencing. Putting flesh in the mouth also suggests that the enemy can be eaten. Cannibalism makes the soldier strong at the attacker's expense.

The skull photograph published in the 1943 *Life Magazine* caused a reader to ask: "'Are we cannibals or headhunters to display the foe's skull ...?" – to which an Army corporal still in training replied that "for the duration" he would be both ... Censors continued to find fingers being mailed home as souvenirs' (Linderman 183). The civilian connects headhunting with cannibalism; the soldier accepts the connection because it provides the kind of desperate strength needed to face war in the Pacific. Extreme measures are naturalized: no one in the war zone asks for explanations, although some soldiers (Sledge's friend Doc) prevent others from losing themselves entirely. After fighting for weeks at the Chosin Reservoir in Korea, a Marine finds himself

> looking around for a place to sit down so I could eat a can of C-rations and I saw the corpse of a Chinese soldier. He was in a kneeling position, resting on his elbows. The top of his head had been blown off. As the body froze, the brain expanded and rose up out of the cranium, looking like a piece of pink coral on a Pacific reef. The hoar frost that fell during the early morning hours covered the top of his brain and was now sparkling in the sun. I opened the can and, using the 'coral' as a sort of centerpiece, sat down on the corpse and wolfed down my meal. (Russ 228)

There's remarkable attention to the detail about the brain, which is touched upon with tenderness, delicacy. The table set with flowers, the sparkle, doesn't in any way deter the soldier but instead delights him and improves his now animal appetite. The dead soldier becomes two household items, a table and fridge. Just as the Japanese skull makes a romantic connection between the 'Navy boyfriend' and the war worker, here the body of the other, when fully domesticated, can almost be eaten. Ingesting the other's body makes for an exhilarating atrocity. A cooked or frozen body makes a legitimate meal, as army slang in Vietnam for napalmed bodies ('crispy critters') indicates. It can even be eaten by one's own people.

Allied soldiers captured by Japanese at the beginning of the Pacific war were put without water or food into the lightless holds of prison ships and transported to work camps. The ships' holds witnessed the kind of soldier atrocities all armies wish to forget. John Keegan suggests that 'inside every army is a crowd struggling to get out' (173), and in the transport ships, where men climbed on heaps of the dead to breathe, that crowd won. One prison-ship inmate remembers surviving by luck, that 'Men were choking each other. Then the awful truth dawned on me as I looked at a body lying beneath me ... His throat had been cut and the blood was being drunk' (quoted in Linderman 221). The soldier body is torn open and the crowd's body, desperate to drink and eat, punches through. No order remains. Army drill cannot withstand the extremes of torture to which the soldiers are subjected. Even in the sternest armies, discipline breaks and the crowd struggles free. Trying to retreat from the eastern front at the end of the Second World War, Guy Sajer and his Wehrmacht friends find the rage to attack German command, famous for its punishment battalions and summary battlefield executions of resisting soldiers. One of Sajer's friends roars at a lieutenant: 'Yes, I'm hungry ... Hungry in a way the saints could never have imagined. I'm hungry, and I'm sick, and I'm afraid, to such a point that I want to live to revenge myself for all mankind. I feel like devouring you, Leutnant. There were cases of cannibalism at Stalingrad, and soon there will be here, too' (384). Religion, humanity, and military spirit are all invoked in this threat to the lieutenant, sign of both the Wehrmacht and the Reich's power. The soldier's desire exceeds the bounds of the body and the war – he is bent on satiating a hunger for revenge as big as the world, a revenge for loss, for the approaching destruction and further shaming of the body. As the army body disintegrates, the soldier, seeking a way to survive, makes a new being if need be by digesting the past self: threatening to eat not only the body but the whole military chain of command points to a future dependent on cannibalism.

New soldiers arriving in the Pacific during the Second World War could use cannibalism as a way to determine humans from enemies: 'On Bataan, a private asked William Dyess, "Cap'n, how will we tell which are Japs and which are monkeys?" Before Dyess could reply, a sergeant answered for him, "Just kill 'em all, son. We can eat the ones that ain't got uniforms on"' (Linderman 169). Beneath the standard racist attack is a message to the soldier body: danger will be eliminated so thoroughly that only bones will be left. What's more, the enemy body isn't worth eating, so the animals will be eaten instead. In Vietnam, Matthew Brennan found himself asking his Vietnamese military guide similar questions:

> 'Con, do you think it's right to call the VC gooks and dinks?'
> He shrugged. 'It makes no difference to me. Everything has a name. Do you think the Americans are the only ones who do that?'
> 'I don't know. I suppose not. I never really thought about it.'
> His eyes glassed over for a moment, and he smiled his hard smile. 'My company in the jungle' – he swept the countryside around Quan Loi with his arm – 'called you Big Hair Monkeys. We kill monkeys, and' – he hesitated for an instant – 'we eat them.' (*Brennan's War* 241)

Although Con has joined the Americans, there is a certain thrill in letting Brennan know that Americans are considered an inferior race, and that they too can be killed and eaten. Now we are at the end of a trail; the body has crawled across the battlefield and has come to rest.

There is a struggle in current military discussions of battlespace over the future of humans, the future of the body. Some argue that it is time for humans to leave battlespace, that they will only clutter a zone dominated by autonomous combat vehicles, self-directed weapons, remotely handled combat. Others claim that no war can be won conclusively without human, particularly infantry, involvement. The strongest supporters of highly technological warfare, first the Navy, and second Air Force, argue that battlespace anatomized by satellites, cruising Unnamed Aerial Vehicles (UAVs), and beams of digital information has no need of humans on the ground or in the air. Conservative military analysts believe that war machines cannot be left unattended. Yet what is the fate of the soldier who remains in battle?

High-tech war running on a digital clock suggests disastrous consequences for the body: 'Man is capable of standing before only a certain amount of terror. To-day there must be swallowed in five minutes what took an hour under Turenne' (du Picq 114). That statement about seventeenth-century Marshal Turenne, Louis XIV's chief of the army, was written by Ardant du Picq in 1868, long before motorized machine-guns, long-range artillery, chemical and nerve gas, napalm, atomic or laser weapons. But du Picq understood the trend: war was speeding up and stuffing more fear more quickly into the body than ever before. Even the conservative Keegan was forced to conclude after Vietnam that 'impersonality, coercion, deliberate cruelty, all deployed on a rising scale, make the fitness of modern man to sustain the stress of battle increasingly doubtful' (325).

Believers in the soldier body follow lines drawn by official U.S. Army historian S.L.A. Marshall's influential although often deliberately misleading book *Men Against Fire* (1947), where he contends: 'In the course of [the Second World War] we learned anew that man is supreme, that it is the soldier who fights who wins battles, that fighting means using a weapon, and that it is the heart of man which controls this use. That lesson we are already at the point of forgetting. We can ill afford it' (23). Marshall's whole career was pinned on his belief in human agency on the battlefield, despite now accepted arguments from psychiatrists that *all* soldiers will certainly go insane when exposed to continuous fire for more than a few weeks. Picking up where Marshall left off is Lieutenant-Colonel Dave Grossman, a military psychologist advertising himself as a 'killologist,' who offers advice about the way 'officers and men can prepare their minds for battle.' Grossman's book *On Killing* (1995), now in its eighth printing, offers itself as a manual for killing to the 'young "virgin" soldier' similar to the sex manuals that offer couples advice on sex – it is Grossman's parallel (xiii). Grossman's advanced-killing Kama Sutra now provides further advice on how to make a *Bullet-Proof Mind* ('Killology'). Grossman's beliefs line up with those of simulation expert Colonel Stephen Tetlow, who agrees with the 'prevailing view of the future battlespace – that is, that conflict is, and will remain, essentially a human activity, in which human qualities of judgement, self-discipline and courage, the moral component of fighting power, will endure ... It is difficult to imagine military operations that will not ultimately be determined through the physical control of people by people. Technology must not be allowed to displace human intent or the decision of a commander' (35–6).

Yet it is difficult, having followed the infant infantry's crawl into industrial battle, to understand how war can ever be about the intimate human endeavour Tetlow envisions. This book turns away from that perspective. I argue that the continued move towards technology produces a new body. Whereas war in post-industrial countries is usually understood to be a practice based on C4ISR, the new war body must come up to speed and combine flesh with armour in order to meet the challenges of virtual battlefields. The first Gulf War and then the 2002 carpet bombing of Afghanistan allowed NATO and NATO-aligned countries to breathe a sigh of relief that the Weinberger (or Powell) doctrine of hitting hard and fast and then getting out produces results. But the lopsided nature of the military powers in those events suggests that technology may not always produce a winner: the 1993 lesson of the disastrous Battle of the Black Sea in Somalia seems to be that the attack of massed flesh on flesh is quite effective. Though the second Gulf War initially appeared to be a textbook case of warfare in battlespace organized by Donald Rumsfeld in accordance with the high-technology mandate of the Revolution in Military Affairs, the persistent and successful rebellions by local militias following the declared end of the ground war (May 2003) have shown, as in Somalia, that armies using the most advanced technology cannot ignore dedicated fighters with old, outdated, or makeshift weapons. Future soldiers' survival will rest in large part on even further technological advances, like powered armour.

What will such body armour consist of? How will it function? The soldier who seeks increasingly sophisticated weaponry will, I argue, begin to lose track of the skin boundary in more subtle ways than I have shown in this chapter. The gun is already part of the body, part of the soldier's defensive skin. Otto Lehrack tells the story of a Marine in Vietnam who seemed to be having trouble with his rifle: 'He put it back up again and started to fire. Nothing. He pulled it back down and was banging it and he kept moving. At the end of the day, come to find out he'd been shot right here [the shoulder] and the round had gone through and through, right through his body, and it had cut whatever nerve it is that runs through your trigger finger. He thought the damned weapon was malfunctioning and didn't know his finger wasn't working' (178). The soldier knows that the finicky M16 isn't to be trusted. But it is his own machine, the trigger finger not the trigger, that is malfunctioning. In battlespace, such confusion between human and metal materiel is one of many to come. The soft body is shocked unbearably by the series of betrayals it suffers in war. Shouldering into an armoured fighting suit seems to be a promising way to avoid the shudder of the mass of flesh.

Chapter Two

Breathing Metal: Armour Suited for War

One tends to see only the thousand tricks of power which are enacted above ground; but these are the least part of it. Underneath day in, day out, is digestion and again digestion. Something alien is seized, cut up into small bits, incorporated into oneself, and assimilated.

– Elias Canetti, Crowds and Power (1960, 1962)

A huge shell thundered: he was vaporized
And, close friends breathing, then internalized.

– R.L. Barth, 'One Way to Carry the Dead' (1994)

Squeeze into some armour: digging into the earth is useless. Throw away that entrenching tool, hurry up. There are gases sifting down, Tabun, Sarin, Soman, mustard, phosgene, a hovering poisonous pharmacopoeia. Back into a clamshell suit, let it close around you and push things into your body; make sure you can move around easily, make sure the relief tubes are hooked up, make sure you can reach the water tube near your mouth, above all make sure you keep a grip. Suddenly it's quiet – there's a cool hiss of already-stale air circulating and the radio channels bring you the fragmentary stutter of other soldiers who, like you, are encased in their private worlds. The weight is on you now, a fully loaded system with the sharpest optics, thermal imagers, Global Positioning Systems (GPS), gyros to keep you level (you're in a big new body after all). Your weapons are targeted off-the-bore, they point wherever you look: you can kill what you see and you can see just about everything. The outside doesn't matter much any more: chemical, biological, atomic, these are not agents that your armour, ticking mechanically to itself as it operates, frets about. The pressures are internal and you bow under them. You only imagine that you feel the bomb racks pressing on your back, the lasers heavy in your mechanical arms. Panic and thought don't fit between you and the suit's inner skin because it's too tight a squeeze. This is heavy duty so straighten up and bear it. Get in it and get over it.

In the early twenty-first century the problem of human flesh at war has become an urgent one. The soldier's mind cannot bear the weight of the flesh's destruction. Soldiers will face unplanned lightning skirmishes just as murderous as conventional fights from the past; the end of the pitched battle promises grim survival rates for armies hoping to stand and fight. Most future warfare will assume the form of so-called Low Intensity Conflicts (LICs), similar to those causing scorched earth in the Balkans, the Caucasus, Palestine, parts of south Asia, and Africa. These conflicts represent what the American military terms either Asymmetric or Fourth Generation Warfare (4GW). The asymmetry refers to a David-and-Goliath contest between a relatively small and under-equipped dedicated band of individuals (like the Chechens, Zapatistas, Palestinians), driven by nationalism, religion, economic hardship (or all of these), with superpowers like the United States or Russia intent on holding land and resources. 4GW is marked by unpredictable, persistent bloodletting. Heavy NATO and Warsaw Pact *matériel* comprising supersonic jets, long-range subsonic bombers like B-52s, tanks, artillery – all is useless against hostage taking, suicide bombing, Improvised Explosive Devices (IEDs), and fierce, fast, evaporating street battles. The Goliaths now find themselves unable to win with the tools at hand (van Creveld *Transformation* 20). The conflict is low intensity only for the absent: as soldiers and civilians often point out – all wars are total wars for the person looking down a rifle barrel. LICs are marked by unpredictability, urban firefights, ambushes, and elusive enemies, as well as all too present swarms of massed hostiles, sometimes civilians, who, through the sheer press of the crowd, overcome powerfully armed contemporary forces ('How many hordes are there in a Chinese platoon?' joked the Marines

retreating from the Chosin Reservoir in 1950 [English and Gudmundsson 148]). An example of Asymmetric Warfare, battlefield
swarming reflects the gaps between the technological haves and have-
nots, between nations pouring more than half their budgets into the
military (as the United States does) and others that assemble ad hoc
militias with particular political or religious purposes.

Technologically heavy forces will continue to fight asymmetric
wars, as the experience in Afghanistan (2001–5) and Iraq (2003–5)
illustrates. But soldiers must, as U. S. Army documents say, be more
'survivable' (able to survive the new battlefield). Despite steaming
debates within the military, the United States Department of Defense
has signed on to the Revolution in Military Affairs, a world-view that
snaps high technology into the machine of fast-moving manoeuvre
warfare. Manoeuvre warfare understands battle as a perpetually shifting affair without the front or rear lines established in previous wars.
The American war in Vietnam saw the beginning of this new kind of
combat, made possible partly because of fresh technology like the
helicopter. Free fire zones, one of the problematic American innova-
tions in Vietnam, took for granted that all humans present in the
zone, no matter their age, gender, or occupation, were to be considered legitimate military targets. People without weapons, including
children and senior villagers, were simply unarmed soldiers. How
can a soldier tell a civilian from a freedom fighter?[6] The RMA's
mode is battlespace dominance through information warfare, particularly using the formula known as 'Command, Control, Communication, Computing, Intelligence, Surveillance, Reconnaissance' – all
subsumed under the earlier mentioned abbreviation, C4ISR (originally C^3I, then C^4I^2). Because the lead words begin with the letter
'C,' all the abbreviations indicate 'C4.' Earlier iterations stood for
the same four Cs and then added 'Intelligence,' or 'Intelligence and
Interoperability,' producing the second superscript I^2 (two Is). The
concepts are the same, however, and C4ISR is often used as short-
hand for other kindred RMA terms. Command and Control (what
officers do in order to reach a military objective) rely on good Communication. In fact, Communication is critical: radios must work,
Internet links must stay up, video must continue to stream. Computing, then, is expected to ensure Communication. In order to be suc-

cessful, the four Cs require quality Intelligence from various sources (humans, machine sensors). That Intelligence in turn depends on Reconnaissance and ongoing Surveillance. Ironically, the compression of the C4ISR acronym includes the entire world of battle – the military seeks absolute knowledge of the battlefield (it must be 'transparent,' as Defense documents say) in order to guarantee victory. Increasingly, RMA proponents look to robots or autonomous machines for the unceasing intelligence gathering and surveillance work, which is known as the 'dull, dirty, and dangerous.'

No matter the abbreviation, the RMA pictures future war as a highly technological activity fought in four-dimensional battlespace where digital computing and communications make it possible for naval, ground, and air forces to exchange (make interoperable) their units. Interoperability is a modular, flexible system where the military objective, not the arm of the military, determines who supports, fights, and commands. But military dissenters argue that the RMA is only a magic act: 'Our ground forces are an attrition force trained to carry out centralized execution of very complete and detailed orders. Despite the trappings of a maneuver force with highly mobile equipment, American military forces are a culture of attrition warfare expertise' (Wilcox 'Maneuver' 157). An overloaded bureaucracy with a top-heavy command structure focused on the stability that comes with heavy technology like tanks, artillery, and long-range bombers negates the advantage of mobile digital technology, argue the RMA's opponents. Foot soldiers will be hindered, their creativity in manoeuvre and problem solving choked off by the long data cables dragging behind the squad. Instead of becoming part of battle's swirling chaos, RMA doubters claim that American troops will operate by sitting in one place and killing everything nearby. Something must happen to free the soldier on the ground, to make the infantry mobile again, to increase the loads they carry, the distances they carry them, and the speed at which all these actions occur.

The solution as proposed by Army doctrine in the publications *Joint Vision 2020* and *Force XXI* is that ground troops must become interoperable units. Soldiers will be connected by networks to the digital battlefield and will appear on the joint commanders' screens as blips to be relocated in cyberspace. In spirit, interoperability means that the task is more important than the force that accomplishes it. If all machines use standardized computing systems and software, it should be possible for a C2 (the officer in charge of the operation) from the Navy (*or* Air Force, Army, Marines) to issue orders for air support whether the jets fly off an aircraft carrier, a Marine tarmac, or out of an Air Force hangar: forces will be interlinked and operate together. As well, they are interoperable in that they become interchangeable, task-based. It is hoped that interoperability will break down traditional communications blockages. Typically, a request from the field will go up one chain of command (Marines), be transferred to another command (Air Force), and travel down that chain until pilots are on the way. These chains of command, or stacks, produce what are called 'information stovepipes': data cannot cross to another force at any but the command level. Interoperability, bypassing typical interservice rivalry, promises to shatter information stovepipes.

The dream of interoperability threatens to render the human a permanently connected creature. What is often called Cyberwar refers to flesh war that has been extended into computer-created territory, depending on *Cyb*ernetic *Org*anisms (Cyborgs) that are fused flesh-metal-plastic beings. Cyborgs, enhanced flesh, spring from Norbert Weiner's term cybernetics, where a sailboat's steersman, sail, and rudder together form a dynamic feedback system that guides the vessel. The Cyborg is a being, like a human with a pacemaker, that operates unconscious of its enhancement. In 1960 Manfred Clynes and Nathan Kline, two scientists invited by the National Aeronautics and Space Administration (NASA) to consider the problem of humans working in space while encumbered by large environment suits, spacecraft and other high technology, coined 'Cyborg.' Worrying about the burdens space technology would press on the body, Clynes and Kline decided that the machine should join with the human: 'The purpose of the Cyborg, as well as his own homeostatic systems, is to provide an organizational system in which such robot-like problems are taken care of automatically and unconsciously, leaving man free to explore, to create, to think, and to feel' (Clynes and Kline 31). The Cyborg is more powerful than either a machine or a human: the combination makes it better. On Earth we usually breathe, see, hear, and move without thought or effort – only when our bodily systems are in trouble do they draw our attention. Clynes and Kline dreamed of the same careless freedom for humans in non-planetary space where there is neither atmosphere nor gravity and the body is attacked by hard solar radiation as well as extreme heat and cold. The body is both liability and possibility in Clynes and Kline's world: it is yet another system that requires operations management for it to function at its peak. Clynes and Kline coined 'Cyborg,' but science fiction had been visualizing humans in space suits for some time, and by 1959 those imaginings were sophisticated, and militarist.

Robert Heinlein's 1959 novel *Starship Troopers* describes in detail a fighting suit that would protect the user against almost all environments, whether deep space, deep sea, or deep enemy territory. The powered suit worn by Heinlein's Mobile Infantry would allow the soldier to fight unhindered: 'You don't have to drive it, fly it, conn it, operate it; you just wear it and it takes orders directly from your muscles and does for you what your muscles are trying to do. This leaves you with your whole mind free to handle your weapons and notice what is going on around you ... the point to all the arrangements is the same: to leave you free to follow your trade, slaughter' (Heinlein 82–3). The human operator works without thinking. Once the soldier learns to use the suit, the armour becomes invisible, its complex features extensions of the human body and self in battlespace. Just as most of us don't consider the intricate commands required to lift an arm, turn a hand, tighten fingers around a tool, so the suit wearer thinks nothing of switching between flame-throwers and rocket and atomic weapon launchers while on a 'smash & run' mission (Heinlein 13–17).[7] At present most soldiers could carry one of these weapons, but certainly they could not carry all three without being overburdened to the point of collapse. The suit permits one to be 'free to explore, to create, to think, and to feel' (Clynes and Kline), 'left free to follow your trade, slaughter' (Heinlein). The uncanny parallel between peace and militarism, the rhetorical twitch from human progress to killing, suggests how close protection is to armour: one skin away. It is probable that Heinlein stayed current with contemporary science; he and Isaac Asimov were both leaders of hard science fiction, defined as SF that uses known or internally

consistent science that obeys the laws of physics and scientific verisi-
militude; he was a lifelong proponent of both the space program and
the military (an Annapolis graduate, he was classified 4F – unfit for
combat duty – just before the Second World War). Whether through
planning or coincidence, Heinlein, Clynes, and Kline arrive at the
same point: a good suit is one for all weathers, including combat's
'storm of steel.' The body evaporates when encased in a powerful
outer shell, and technology's logic of extension takes over: the Space
Cyborg becomes a War Cyborg.

Despite the existence of complex environment suits that allow
astronauts to walk in space for long periods, there's still a gap
between science fiction's waking dreams and nightmares. The Army
proposed a 'supersuit' that its new soldier – the 1991 iteration was
called the 'Land Warrior' – might wear (see fig. 3); the suit would
comprise 'light-weight ballistic armored material'; extra underbody
armour for defence against mines; an infrared scrambler; 'a built-in
cooling system in which a coolant substance flowed through the
outer skin via a network of plastic capillaries,' erasing any thermal
signature and thus making it invisible to infrared scopes; and liquid
crystals woven into the suit's fabric that would mimic 'the color of
the underbrush; in desert, the color of sand, rocks, and arid vegeta-
tion' (Alexander 95). The Army hastens to point out that the Land
Warrior 'is a fighting system' where the human is the ignition key
(Garamone 'Army Tests'); it is a system that erases the body's
vapour trails and matches the human to a war environment. The body
is partly armoured, partly invisible; the other materials (thermal
imaging cameras, zoom lenses, GPS) enhance human senses and
allow network sensory exchanges where helmet mounted displays
allow the soldier to e-mail the squad with maps and positions, even
screen shots, of what the rifle sees (Jontz 1). The human head is now
a mount for the Integrated Helmet Assembly Subsystem (IHAS),
which is connected to the rifle, camera, and GPS by a long umbilical.

The umbilical is an intriguing sign of the human entangled in a
newly created machine. There is still more human than machine, but
the subversion has begun. The mechanical lifeline on the soldier's
chest indicates how crude the wearable computer is. What's more,
we're still talking about material, clothing shot through with some

3   **New suits**: The Land Warrior with early, bulky, thermal, video sensory pack on rifle; a data cable connects to a wearable computer in the soldier's vest.

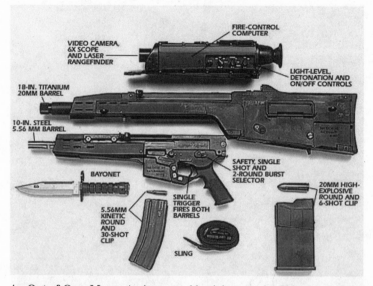

4   **Out of One, Many:** An interoperable nightmare, the Objective Individual Combat Weapon (OICW), with its modularity highlighted in this view. The three main weapon sections include the streamlined video-computer pack (the fire-control unit and computer also acts as the weapon's sight); the high-explosive 20 mm round grenade-launcher equivalent; and the smaller automatic rifle, which fires standard NATO 5.56 mm 'kinetic' rounds (non-programmable bullets that damage flesh by tearing it apart the old-fashioned way). The gun is 'mated' in an 'over and under' manner (the heavy partner, the grenade launcher, is on top). Note that a bayonet has been included.

smart fibres. Operating on the premise that what is within reach is already obsolete, and as the Land Warrior enters its second year of testing, the Army has moved on to its new infantry model called the Objective Force Warrior (to be replaced by the Future Force Warrior by 2010).

The Objective Force Warrior (OFW) comes with a full set of accessories, one of which is the Objective Individual Combat Weapon (OICW) – it's a gun (see fig. 4). The OICW is also sometimes called Selectable Assault Battle Rifle (SABR) – the acronym gears are grinding. In itself the OICW is a kit acting as a remote control for digital war: it sees and fires around corners; throws kinetic and 'bursting' ammunition, a capability that eliminates the standard M203 grenade launcher clipped to the M16a2 rifle barrel; calculates windage and range, then programs the chip in the exploding round to detonate at the right time and position. The OICW is a picture of modularity and interoperability. At 5.5 kg, it is light and breaks into two pieces so that unburdened ground troops can become even more fleet by disposing of the top or 'over' part of the gun. The weapon is a relentlessly masculine sexualized dream that mounts the heavy launcher on the lightweight automatic rifle. Gunsmith language distinguishes the 'over and under' quality of the rifle and grenade launcher, eroticizing a weapon that is always engaged in sexual intercourse, or at least domination, unless one of the partners is discarded. The gun talks to itself, its ammunition, the soldier, the target, other squad members, and any Command and Control personnel who want to observe through the riflecam. The weapon, in full communication with its close and distant environments, is a node in the network of RMA interoperability.

Even the Objective Force Warrior, the 'soldier of tomorrow,' a character in the Combatland theme park, is a small step along the evolutionary path of the human in battlespace. The Revolution in Military Affairs will not stop with some smart materials and helmet-mounted sighting systems; it's a new body that is required, one that conceals flesh, enhances physical power, strengthens the self against battle-shock, and reinforces the fragile ecosystem of a shuddering body that must be reconfigured to terrify the enemy even as it reassures the soldier.

The infantry soldier's load is limited by the physics of the human frame. The Army would prefer it if the soldier didn't carry more than a third of the body's weight into battle, but this ideal, even given constant air resupply of water, food, ammunition, medicine, is hard to achieve. Reorienting itself, the Army decided that rather than lighten the pack, it would strengthen the soldier (now called a 'platform'): 'For the first time, the soldier's equipment is being designed as if he is an individual complete weapons platform. Each subsystem and component is designed to and for the soldier. The result: the first integrated soldier fighting system for the dismounted infantryman' (FAS 'Land Warrior' 1). The soldier – metonymically a 'dismount' – rides to war in a LAV or helicopter, then steps down and fights. But perhaps it isn't a soldier who dismounts. Instead of a body loaded with gear that alights, it might be an engineered weapons platform. Sarcos, a robotics company, has worked on an exoskeleton that mechanically bulks up the human (see fig. 5). Working models of similar systems have been built at the University of California at Berkeley's Human Engineering lab. The Berkeley Lower Extremity Exoskeleton (BLEEX) is a 45 kg armature that makes a 32 kg pack feel like a 2.25 kg load. Videos of the BLEEX show the human moving slowly but purposefully under an enormous pack as the machine puffs away like a hydraulic monster. In case the viewer feels secure at the mechanical distance the machine has to go, it's worth noting that only two decades separated the First World War's ungainly tanks from the nearly unbeatable Second World War German PzKwVI 'King' Tiger tank. The more the human puts on, the less weight the body feels. The cyborg, a homeostatic killing machine, has arrived at the edge of battle where flesh and machine operate in such close proximity that they seem almost inseparable. There's no need to bother with cloth uniforms if a new outer skin can take an electric charge and change colour. The body betrays the soldier from within: hit by a kinetic charge, the bones suddenly become internal shrapnel, jagged knives cutting open flesh. It's time to force the body into a rigid shell like the one Robert Heinlein imagines:

> A suit isn't a space suit – although it can serve as one. It is not primarily armor – although the Knights of the Round Table were not armored as well as we are. It isn't a tank – but a single M.I. [Mobile Infantry] private could take on a squadron of those things and knock them off unassisted if anybody was silly enough to put tanks against M.I. A suit is not a ship but it can fly, a little – on the other hand neither space ships nor atmosphere craft can fight against a man in a suit except by saturation bombing of the area he is in (like burning down a house to get one flea!). Contrariwise we can do many things that no ship – air, submersible, or space – can do. (80)

Heinlein defines the suit by a process of exclusion, a rhetorical device underlining the suit's capabilities: it isn't any one thing but

5 **Bearing It**: The Sarcos exoskeleton is a lower-body brace with a waist-level power pack. Ultimately it will also enclose the shoulders and hands, enabling the user to shift vehicles and other large objects.

excels at all things. In swallowing the human, the suit seems to render all but total war obsolete. Heinlein's suits, worn by the Mobile Infantry, Leathernecks in hardware, are a late-1950s dream of masculinity, of men without women. In Heinlein's world, men of honour prove themselves by fighting on the ground but the women, who are finer, rarer creatures, make the best starship pilots. The division of labour separates clean from dirty killing: men in suits are delivered to the blue-collar labour of battle by women in their floating ships: while there seems to be equality in battle, in fact it is only the men who have control over their personal bodies and body armour. By 2002, in order to make the Objective Force Warrior (OFW) technically feasible, the Defence Advanced Research Projects Agency (DARPA) awarded contracts for an agency to build an infantry exoskeleton and create the Institute for Soldier Nanotechnology.[8] The competition winner, the Massachusetts Institute of Technology (MIT), fantasizes this way about the future warrior: 'An "exoskeleton" for the soldier composed of such things as novel nanoparticles, electroreological fluids, and polymer actuators could not only provide ballistic protection, but also be transformed into a medical cast (on demand). Alternatively, it could be activated to create an offensive "forearm karate glove"' (MIT *News* 2). As with the OICW modular combat weapon, the key to MIT's nanomachines is their mutability. The material is a permanent fluid, always about to become something else. The weapons of the Revolution in Military Affairs are chameleons, adaptable, all too agile. Future armour will soften, harden, administer medicine, and monitor the body's performance. Human bones will be reinforced by DARPA's exoskeleton, a new barrier against combat. No longer need the soldier fear the terrible intrusion into flesh that the rain of battlefield metal presents: flesh need not crawl if the enhanced body can leap away. While initial plans for the exoskeleton are cumbersome and limited, there's a logic of machinery at work that will see flesh retreat as nanomuscle advances, building the new soldier body.

Back from a stint as a combat engineer in Vietnam and recovering from wounds suffered during a massive explosion that shredded a good deal of his body and all his trust in the military, hard science fiction author Joe Haldeman reconsiders *Starship Troopers*' Mobile

Infantry. Haldeman's *The Forever War* (1975) contains a grim revision of Heinlein's powered armour: whereas Heinlein declines to give specifics about the suit except to indicate how well it worked, Haldeman concentrates on the way humans use and abuse machines:

> 'I assume you know how easily the waldo capabilities can kill you or your companions. Anybody want to shake hands with the sergeant?' He paused, then stepped over and clasped his glove. 'He's had lots of practice. Until *you* have, be extremely careful. You might scratch an itch and wind up breaking your back. Remember, semi–logarithmic response: two pounds' pressure exerts five pounds' force; three pounds' gives ten; four pounds' twenty-three; five pounds', forty-seven. Most of you can muster up a grip of well over a hundred pounds. Theoretically, you could rip a steel girder in two with that, amplified. Actually, you'd destroy the material of your gloves and, at least on Charon, die very quickly. It'd be a race between decompression and flash-freezing. You'd die no matter which won. (*War* 17)

Haldeman's armour is identical to Heinlein's in that it vastly multiplies the body's power. But Haldeman, a veteran foot soldier, won't forget how stupidly terrified people act, whether or not they're locked into death machines. He reminds the reader that 'most of the things that are truly effective against the enemy tend to be effective against the user as well, if he trips or sneezes or forgets to put on the safety before he scratches his ear with the muzzle. (You think that doesn't happen?)' (*Armor* 4). Haldeman's sardonic parenthetic question undercuts the boy scout quality of Heinlein's eager Mobile Infantryman Johnny Rico, who believes that fighting and dying is more valuable to the state than any other activity the citizen can perform. *The Forever War* indicts the suit as the hand of military and state power clamped onto the human, without looking away from the armour's still impressive physical qualities. The universe is a deadly place, says Haldeman, but being inside the suit doesn't promise more safety; the stuff will rupture, guaranteeing the body's immediate destruction. There is no armour against clumsiness, forgetfulness, rage, or the need to scratch an itch. The body is immured, hands tied down to other tasks dictated by the armour.

MIT has begun working on the first phase of DARPA's exoskeleton project. Recently the team reported on materials with surprising or unpredictable qualities. Current technology doesn't seem to have any downside to it: it will increase a soldier's carrying load, running speed, and lifting ability. But the final suit may be a good deal more disturbingly pliable, its boundaries less distinct: 'That increase in how much a polymer can expand and contract, combined with its impressive strength – which the researchers haven't measured yet but predict to be ten times that of human muscle – could conceivably allow a combat uniform embedded with 1.4 kilograms of material to lift 80 kilograms one meter high. In other words, a soldier could effortlessly hoist a heavy piece of equipment or even a fallen comrade' (Talbot 'Super Soldiers'). MIT's Institute for Soldier Nanotechnology looks for unusual, surprising technology reasoning that if we've never seen such objects, they won't yet be obsolete. The technology that DARPA is currently field testing in the Objective Force Warrior program is old. New armour is going to be disturbing because we won't recognize it: perhaps the polymers that can perform with such force will be implanted directly into the body. Perhaps the new suit will revise the body's flesh itself. Even Heinlein understood that soldiers were part of the machine creature, small metal seeds ejaculated out of huge steel pods: '*Bump!* And your capsule jerks ahead one place – *bump!* And it jerks again, precisely like cartridges feeding into the chamber of an old-style automatic weapon. Well, that's just what we were ... only the barrels of the gun were twin launching tubes built into a spaceship troop carrier and each cartridge was a capsule big enough (just barely) to hold an infantryman with all field equipment' (9).

Heinlein's reassuring image of the automatic, the quintessential slick modernist Second World War sidearm naturalizes the idea that humans are bullets. The M9 Beretta 9 mm handgun is part of war comfort: we're used to automatic weapons, central to the National Rifle Association – revised holy writ of the U.S. Constitution's second amendment. Few weapons have as much contemporary mythic significance for an action-loving machine culture as the automatic pistol or pump-action shot-gun – only Clint Eastwood's Dirty Harry Callahan managed to keep the revolver in vogue. Powered armour, no matter either Heinlein or Haldeman's intentions, looks attractive. It suggests that, even in the chaos of battlespace, one can control the body and affect the outcome of combat. In promising new power and continued life to the human, Cyborg machinery appears to be benign. Accepting these additions, however, will lead to others that are perhaps less palatable. The BLEEX is reassuringly mechanical: it's an addition to the human frame and doesn't involve injections, insertions beneath the skin, remaking the body. The boundary of human and machine seems quite evident at the moment. The lines are clear. The erosion of the skin, the gradual sloughing off of the skin – that is to come.

The armoured soldier may not be wearing anything identifiable or reassuring: the slightly bulky-looking battledress that shifts and glints in the light, bulging over the shoulder and around the upper arms and calf muscles where prosthetics have crept in, will make us uneasy. The body will have altered its shape and signature. When we look into a thermal imaging sight or infrared scope, it will be hard to see the new soldier. Heartbeat, sweat, odours, waste heat, all these things will have been made to vanish. Nanomachines busy remaking the body's effluent will render the human into a sort of shadow on the scope. Blurred hands wrapped in strange materials will lift and throw bodies, gut them, or smash them into lumps of shattered bones and ruptured tissue. Haldeman and DARPA's Institute for Solider Nanotechnology imagine a future darker than Heinlein's. It will be a place of utter terror with anxious near-insane users, and no reassuring past technology to comfort us.

Humans have been washed up into, driven through, and dropped on the battlefield. Because America's Joint Forces are more focused than ever on the littoral environment, as the Navy's 1992 vision statement 'Forward ... From the Sea' indicates, battle will be carried by troops moving from the ocean (troop or aircraft carriers) onto land. Possession of offshore waters obviates the need for awkward political alliances. Ground assaults will be launched from the ocean and the air. The reason the Army can buy into the RMA is that, even in battlespace, someone must land and fight, which is what 'makes the airborne infantry useful and unique. They arrive in bulk, right from America. Unlike any other conventional unit shipped from the U.S.

homeland, the paratroopers can fight their way in' (Bolger 64). The infantry, whether constructed as jump troops (Airborne), mounted troops (mechanized), air assault (heliborne), or ocean-going (Marines), form (in Bolger's pugnacious rhetoric) a breathing, hot export commodity. Bolger uses the language of markets to discuss warfare; to secure an economy, export troops. As with Heinlein's picture of soldier-bullets, humans are products that must survive long enough to accomplish their tasks; shore environments are murderous for unprotected troops, as every soldier involved in a beachhead assault remembers. No wonder the Army dreams about space as the new kill zone; there an orbiting vehicle would release clouds of armoured soldiers who would, unlike relatively helpless paratroopers, perform armoured suit-powered dives through the atmosphere to combat. The new margin is that which separates a high-atmosphere parachute drop from a space-borne attack (the latter is Heinlein's ideal). But to make such a high fall possible, the soldier must be wearing an immaculate, expensive, rig: 'I just want to remind you apes that each and every one of you has cost the gov'ment, counting weapons, armor, ammo, instrumentation, and training, everything, including the way you overeat – has cost, on the hoof, better'n half a million. Add in the thirty cents you are actually worth and that runs to quite a sum.' He glared at us. 'So bring it back! We can spare you, but we can't spare that fancy suit you're wearing. I don't want any heroes in this outfit' (Heinlein 6).

While the sergeant covers his love for his boys by complaining about the suit costs, those costs are a valid concern. Personal armour is going to be as expensive as contemporary attack helicopters (currently near the $30-million-USD mark). The Army's investment in pricey hardware is something for opponents of the technologically driven RMA to scream about. Indeed, the Army and the legislative and executive branches of government have demonstrated more care for machines than humans, as the fate of veterans of two Gulf Wars shows. Soliciting new and replacing old equipment is the main task of the organization now identified as the Military Industrial Congressional Complex (Richards 297). Humans cost more than the sergeant's dismissive few dimes, but, relative to the machinery, the sergeant has the numbers about right.

Still, if even cheap and expendable infantry are going to survive battlespace, armour appears to make sense. Battlespace is a big open translucent murderous machine: 'Battlespace is that volume determined by the maximum capabilities of a unit to acquire and engage the enemy – capabilities that will be greatly expanded by future technology' (*Force XXI* 3–8). The RMA pins its entire doctrine on the battlespace transparency conferred by what is called informational awareness. Information warfare is intended to render battle wholly visible, removing Prussian militarist Carl von Clausewitz's famous friction of war. Instead of a battle*field* (a plane), *Force XXI* shifts to 'volume,' a three-dimensional zone that exists across time. Controlling and commanding in that volume requires not only hovering drones and Unmanned Aerial Vehicles (UAVs), aircraft like the Joint Surveillance Target Attack Radar System (JSTARS), whose job is to report on ground activities, and Airborne Warning and Control System (AWACS), which looks for air-based threats, but also a powerful Navy drifting in the shallows to provide resupply, air support, and 'reach-back' weapons (one example is artillery so powerful that it can 'reach back' to the most distant of the enemy's lines – in some cases more than thirty kilometres). Long-range bombers will troll fifteen or more kilometres overhead, out of vision but well within bombsight. Into this open fluid of digital awareness, where networks e-mail each other screen shots from the front, views from craft above or robots on the ground, lumbers the armoured suit with a human inside. Just as air cavalry excited a generation of commanders in Vietnam, so DARPA enthusiastically looks to personal mobility. Trek Aerospace built and tested the SoloTrek XFV (fig. 6), and has now moved on to the Springtail EFV, both vertical take-off and landing personal aircraft (somewhere between an airframe and a flying forklift): '*SoloTrek XFV* (Exoskeleton Flying Vehicle) is a brand new kind of transportation device that you step on, strap on, and fly. This ultra-compact aircraft lets you takeoff vertically, dash to your destination, then land literally anywhere' (www.solotrek.com/mjet/index1.html). The key words are about speed and Cyborg carelessness. Trek Aerospace insists that little or no training will be required in order for a soldier to operate the craft. The new body will be able to operate without thought, or, in less euphemistic language, it will be launched at some velocity into an

6   **Above It All**: The SoloTrek XFV (Exoskeleton Flying Vehicle) looks as ungainly and unlikely as the Wright brothers' first gliders. It's a start, however, into the world of personal combat mobility.

open battlespace where high technology, having greased war's friction, has created a glassy place of death. If the machines fail there will be little hope for humans now footless on that slick plane.

It's a short slip from Solotrek to Haldeman's science-fiction visions about the fate of a damaged soldier imprisoned in a wrecked machine: he holds 'the smoldering stump of his arm, seared off just below the elbow. Blood sprayed through his fingers, and the suit; its camouflage circuits scrambled, flickered black-white-jungle-desert-green-gray' (*War* 69). The suit's final thoughts show on its surface as it tries to match an environment of death. None of the camouflage settings will save it as it thrashes to ruin, blood and circuits scrambled together. When the medics come, the armourer must follow; sometimes the armourer must come first. Because the war Haldeman imagines goes on for more than a millennium, the armour designers improve on their early models. The suits get smarter about the body: 'The suit is set up to save as much of your body as possible. If you lose part of an arm or a leg, one of sixteen razor-sharp irises closes around your limb with the force of a hydraulic press, snipping it off neatly and sealing the suit before you can die of explosive decompression. Then "trauma maintenance" cauterizes the stump, replaces lost blood, and fills you full of happy-juice and No-shock. So you will either die happy or, if your comrades go on to win the battle, eventually be carried back up to the ship's aid station' (*War* 158). The suit is a hard shell, a place of refuge, a weapons platform, a 'force magnifier' (as the Army calls such devices), and also a small factory. The body is closed into a stamping mill that can sever limbs with hydraulic force and machine speed. The factory operates with such celerity that it can beat the invasion of vacuum, efficiently clean up the remaining product, shut down the physiology, and put time and emotion on hold. If there is no hope for the encased body, then the industrial coffin itself becomes a combined crematorium and weapon. When one of the soldiers is killed the sergeant tells the troops:

> 'She'll be taken care of. From the ship.'
> After we'd gone half a klick, there was a flash and rolling thunder.
> Where Ho had been came a wispy luminous mushroom cloud boiling
> up to disappear against the gray sky. (Haldeman *War* 54)

The soldier has completely forfeited the body, which doesn't just become waste, or even bone shrapnel, but a radioactive cloud that will go on killing those around it even if they've missed the first blast. Armour has encased flesh and gets busy cleansing itself of its human pollutant. The body is only a binary flicker away from atomic destruction. The suit hangs between operating as a protective force and as an agent of imprisonment, a place of incarceration where the state has full command over the body's future, which will either be chunks or radioactive vapour. Armour is on the rise, getting stronger even as the body continues to disappear through a series of invasions, flayings, skin that no longer successfully protects an interior.

The body has a last few chances to show itself. Haldeman contrasts the armoured and the exposed: 'After a lot of shuffling around, every-body finally got plugged in and we were allowed to unsuit – ninety-seven naked chickens squirming out of bright green eggs' (*War* 21). Bright green eggshells are the new wombs of creation, matrices of protection which would, were the soldiers to leave them, cause death. Being born from armour means being pushed screaming out into battlespace entirely vulnerable: birth means immediate destruction. Soldier-chickens squeezing out of tight suits, form-fitting armour that picks up its impulses from the host's flesh and nerves, are revealed in their shame. If being human rests on preserving emotion-producing flesh, then humanity cannot be saved by muscular strength. Flesh, even a mass of flesh such as that which so horrifies Elias Canetti, can-not stand against armour; inside the suit the body has already lost, out-side the suit bodies are about to lose. Haldeman's soldiers step into their suits and the fronts swing shut: 'I finished the set-up sequence and the suit closed by itself. Gritted my teeth against the pain that never came, when the internal sensors and fluid tubes poked into your body. Conditioned neural bypass, so you felt only a slight puzzling dislocation. Rather than the death of a thousand cuts' (*War* 156).

All the soldier can do, all that is left, is to grit the teeth. Armour has worked its way around, clamped down onto, and finally pushed under the skin. Even the skull, that redoubtable internal helmet, hasn't saved the brain from a rewiring, reducing the sense of bodily intrusion to a minimum. Without armour, the body is a fearfully soft object, a bag of tender contents that can be ripped open by enemy fire; inside the suit, the body is vulnerable to penetration by medical instruments. The dreadful convergence of medical and military tech-nology makes the body into a horror zone: the skin is too large to be protected. Haldeman's soldier-narrator warns that it 'takes a lot of hypno-conditioning to lie there and have oxygenated fluorocarbon forced into every natural body orifice and one artificial one. I fingered the valve embedded above my hipbone' (*War* 106). The armoured suit requires that it implant itself in the body; if the body is going to survive, the machine must grow into it. Star travel and the technology of speed requires that the body be held immobile in oxygenated fluorocarbon: new orifices further expose it to invasion. The human being recoils from what the body becomes because it's possible that these modifications will cause a despair too deep to be borne – it's possible that the person, not just the flesh, has been insupportably discovered.

As armour becomes more penetrating, the elusive self in its tricky, risky body container retreats. Humans compressed during traffic accidents into bone-breaking shapes find enough to panic about without exchanging the highway for the battlefield. Technological insertions of the kind we can expect to see in the new decade will be more intimate than ever, more likely to cause the soldier full-blown terror. The armoured soldier will need to find a home for the self that is outside the body: there will be little room left inside. Anthropologist Mary Douglas explores existence on the edge of the self, on the brink of deep bodily contamination: 'All margins are dangerous ... Any structure of ideas is vulnerable at its margins. We should expect the orifices of the body to symbolize its specially vulnerable points. The mistake is to treat bodily margins in isolation from all other margins' (121). To understand only the skin as a boundary is to forget that the body is a series of layers, wrappings of ideas, flexible sheets, protections. The shame of the exploded, fractured, wounded, body revealed in combat arrives with catastrophic speed and requires an infinity of healing time. Peeled open, the body leaves the entrails of human identity hanging out. Confronting an armoured suit, the soldier recognizes the *whole* body as a marginal zone, an understanding that makes the soldier queasy about all technology since what were once external mechanical artifacts are now potential boundary breakers. Two of Haldeman's armoured soldiers try to comfort each other, but the nightmare only intensifies: 'I put my hand on her knee. The contact had a plastic click and I jerked it back, visions of machines embracing, copulating' (*War* 55). The soldier sees humanity as a mirage beyond the view plate – only digital contacts remain. Armour is the new power. Armour is the agent of the human disappearance. Armour begins to live.

Turning from the future to the present, weapons and weapon systems obey their own logic. Even a small weapon, if it is influential enough, can become the centre of a military galaxy. Looking back won't brace us for the next iteration of bloodletting (preparing for the next war based on what had already occurred stalled the French in the conceptual trench of the Maginot line), but it can give us hints about how weapons become constellations in themselves. Military historian Martin van Creveld argues that 'to beat one technological system, it is necessary to direct against it another either much more powerful or much more flexible' (*Technology* 195). It seems that asymmetric warfare counters military organization with chaos; but chaos is itself a practice, and technology, whether box-cutters or armoured suits, are identifiable structures (in an example of Cyborgs at work, box-cutters connected humans to civilian aircraft, transforming them into explosive flying whales on 11 September 2001). Technology calls out to technology, system brings on system.

The machine-gun, a weapon often claimed to have frozen the First World War battlefield in Hiram Maxim's sights, is an example of weapon genealogy, logic, and assemblage. The machine-gun produced a particular kind of infantry war where, 'by 1918, an impartial observer might even go so far as to say that the infantry consisted of two types of troops – those who served machine guns and those who specialized in destroying them' (English and Gudmundsson 15). Each side had to use the machine-gun because the other did; the gun had to be destroyed since, while it operated, no movement on the battlefield was possible. The gun's swift design evolution during the First World War led to a portable, accurate, reliable, powerful killer. Its reach emptied out the battlefield, not simply by producing hundreds of thousands of corpses (the gun is often referred to as a scythe or reaper, and for a while the light machine-gun was, in a quaint domestic fashion, called a 'trench broom,' sweeping away bodies like leaves), but because it could give users control over hundreds of metres of the front. Squads dispersed, soldiers strung out so as not to be caught in clusters. Rather than look for the enemy, soldiers became alert for bullets hitting around them; visible enemies meant probable death for both parties. The empty battlefield, a shock to commanders accustomed to late-nineteenth-century close-order fighting, was part of modernity's high-speed, long-range war.[9] The newly created twentieth-century soldier was really two or three soldiers: the rifleman now killed beside people handling crew-served weapons like machine-guns, mortars, small artillery pieces – weapons requiring two- or three-person teams to operate. Two words are of particular significance: the 'crew' has so far been reserved for artillery but now comes into the forward line because one person is insufficient to manage the force and complexity of new killing devices; more important the word 'served' suggests that humans have been pushed behind machines, that 'Through bitter experience the machine taught that man himself was no longer master of the battlefield. The individual counted for nothing, all that mattered now was the machinery of war' (Ellis *Gun* 142). The machine-gun, 'Fordized' factory death, requires a team to serve its assembly-line slaughter.

The machine-gun wasn't new in the First World War. Types of machine-guns were used in the Crimean War, the American Civil War as well as in numerous British colonial massacres. The gun was emergent technology brought to full force by economics, politics, terrain, war-production methods, and profiteers. It was familiar but also suddenly new. The appearance of powered armour in the American Cold War technological and cultural imagination suggests that, as DARPA's machine dreams – Haldeman's nightmares – approach feasibility, armour will arrive on the battlefield and dominate it as the machine-gun once did. When it arrives, armour will meet armour as well as light portable hit-and-run weapons whose sole aim is armour's destruction. Armour-killing weapons already exist as Fuel Air Explosives (FAEs), known in their shoulder-fired incarnation as thermobaric grenades that produce 'a tremendously powerful volumetric explosion that is deadly to troops in bunkers, buildings, foxholes, around corners, or in light armoured vehicles. Body armour actually enhances the effects of this weapon' (Sayen 183). FAEs have been used by the Russians in their recent Chechen wars and by the Americans in Afghanistan; they are known as 'non-nuclear

nuclear weapons.'[10] They kill by fire and vacuum, causing over-pressures in the body, rupturing lungs, eyeballs, and other soft tissues, sucking air out of enclosed spaces like bunkers or bomb-shelters (armour acts like a small bunker). Because armour is expensive, most countries will invest in armour-killing devices like FAEs. Battle will become an insane scramble between armoured and exposed soldiers carrying shoulder-fired thermobaric weapons. As long as the supply of battlefield gamblers isn't exhausted, FAEs will look like a bargain.

Armour, like most developing and desired high technology (incendiary weapons were considered a waste of effort until it came time to kill whole cities like Dresden, Hamburg, and Tokyo), creates a familiar logic of speed and space, requiring extensive technical supply and support umbilicals that attach to and maintain the armour. Armour demands of its enemies small, fast, cheap technology, and preparation for battlefield suicide. Armoured soldiers capable of actions performed by helicopters, tanks, and LAVs, will understand their bodies in new ways. The soldier must become a body pilot accustomed to living a simulated life; in fact, simulation *is* life for the armoured soldier, as it is for most pilots. Cultural theorist Mark Seltzer proposes a panic-exhilaration continuum where the panic of being closed into armour is matched by a sense of invulnerability that propels the soldier into actions an unarmoured human wouldn't consider. Armour won't make people brave: it will make people we don't recognize. Armour-schooled soldiers will view with some irony the frail human body as a liability. Being born is to be armour-borne.

Haldeman's *The Forever War* (1975) was in part a response to the dangers the body faces in war. His philosophical sequel *Forever Peace* (1997) removes human bodies, if not minds, from the battlefield entirely. He proposes a system of remotely operated ground ('soldierboys'), air ('flyboys'), and marine ('waterboys') fighting machines. Platoons of soldierboys are directed by human 'mechanics' closed into shells, connected to each other by neural nets that force them to think, sense, and live through each other (the narrator speaks of 'one creature with twenty arms and legs, with ten brains, with five vaginas and five penises' [Haldeman *Peace* 8]). The collective body is stored safely in the continental United States; flesh is sidelined while soldierboys drop down to the planetary surface to fight. What's left of the flesh soldiers are called '"shoes" [who] always faced a certain probability of being ordered to go out and sit on a piece of disputed real estate. They usually didn't have to fight, since the soldierboys were better at it and couldn't be killed, but there was no doubt that the shoes fulfilled a valuable military function: they were hostages' (46). Armour has wrapped around and pressed into the body until finally the body is abolished. Just as system summons system, flesh hails flesh: a hostage soldier is a Judas goat that politicians can use to shepherd the country into military action for oil, water, or other equally desirable resources.

As flesh becomes an increasing problem, the Army must set armour development at full throttle; something must come to save the tissue of the self. Heinlein's Mobile Infantry literally bounce around calmly throwing tactical nuclear weapons because armoured soldiers can commit atrocities against flesh – armour invites such acts. A mechanic safely operating a soldierboy thousands of kilometres away, as full of adrenaline and murder as if on the field, produces dreadful results for the flesh: 'Park grabbed an ankle and swung the man like a doll, spattering his brains on a concrete slab, and tossed his twitching body into the mob. Then he waded into the crowd like an insane mechanical monster, kicking and punching people to death. That snapped me out of my shock. When he wouldn't respond to shouted orders, I asked Command to deactivate him. He killed more than a dozen before they complied, and his suddenly inert soldierboy went down under a pile of enraged people, pounding it with rocks' (Haldeman *Peace* 137). The supersoldier, no longer a body but a walking exoskeleton, a true weapons platform, is buried in a mass of flesh humans. The swarm, so fearsome and effective, is a sign of people's loathing for apparently autonomous technological forces. Technology is suddenly all too lively, unstoppable, the body a helpless, impotent rag, and the only weapon to hand is the oldest: a rock. The machine is infused with human emotion and action but Command's world is all business euphemism. Enemies prepared to die by the thousands (the Chosin Reservoir campaign 1950, Mogadishu 1993) or hundreds of thousands (Vietnam 1961–75) in the face of high technology seem impossible to beat without their total annihilation. Only truly heavy technology like tanks and high or Fuel Air Explosives slow down a swarm – it was armour and personnel borrowed from the United Nations Operation Somalia (UNOSOM II) that saved American troops in Mogadishu. The soldierboy can be stopped only by a corporate-ordered lobotomy: it isn't a classic story of a metal creation gone berserk but the picture of a human disconnected from an insensate body.

Haldeman's soldierboy is in the works. Confronted by the maze of the body's soft complexity, scientists have begun to design new flesh. In battle, liquidation of the other's flesh signifies victory: it follows that ensuring success will come from creating indestructible flesh for the self. Research at the Institute for Soldier Nanotechnology promises that 'Applications are countless. According to Yoseph Bar-Cohen, senior research scientist at NASA and author of *Electroactive Polymer Actuators as Artificial Muscles* ... "This is an exciting field. We can start thinking in terms of copying nature. Instead of having motors with gears and bearings, we'll soon be able to take a blob, attach wires, and it will change shape as desired"' (Cameron 'Artificial Muscles'). A malleable blob is a better candidate than the complex structure of the human body for absorbing battlefield ferocity. But some force must shape that blob, squeeze it to fit the available battlespace, and DARPA is preparing that force, too. DARPA's 2001 contract for research into Brain Machine Interfaces (BMIs) has been awarded to a consortium of scientists at Duke University. The reason for DARPA's interest in BMIs is clear when we examine their definition: 'The brain takes inputs and generates outputs through the electrical activity of neurons. DARPA is interested in creating new technology for augmenting human performance through the ability to non-invasively access these codes in the brain in real time and integrate them into peripheral device or system operations' ('Brain Machine Interfaces'). DARPA understands the brain as a digital encoder operating in the here and now. It would like to tap that code without surgery and send the commands to remote objects like artificial limbs – or robots doing various tasks. Duke politely focuses on the prosthetic-limb applications for which there is a projected 1.5 billion-USD market by 2005 (Duke 'DARPA'). By early 2004, the Duke team reported significant advances in training monkeys, then humans, to control robotic limbs using only brain impulses. Neuroprosthetic devices, as they're called, are governed by deep subcortical electrodes that connect wirelessly to a computer and then the robot in question. DARPA wants more than prosthetic limbs (the 'D' stands for 'Defense,' after all); they desire human brain-controlled robots capable of 'complex motor or sensory activity (e.g. reaching, grasping, manipulating, running, walking, kicking, digging, hearing, seeing, tactile). Accessing sensory activity directly could result in the ability to monitor or transmit communications by the brain (visual, auditory, other)' ('Brain Machine Interfaces'). The images Duke uses show Rhesus monkeys completing a task using a joystick and video screen. After a number of repetitions, the joystick disappears yet the task continues. Physical motion dies off: 'Her arm muscles went completely quiet, she kept the arm at her side and she controlled the robot arm using only her brain and visual feedback. Our analyses of the brain signals showed that the animal learned to assimilate the robot arm into her brain as if it was her own arm' ('Monkeys'). The body becomes spectral when what philosopher Gilbert Ryle called the 'ghost in the machine' leaps from the brain to animate the robot – there is little separation visible if the monkey understands the arm has become superfluous. The brain is DARPA's true tissue, to be kept remote and protected as Haldeman predicts. DARPA will be at the forefront of the lobe wars, writing 'new algorithms' that will read-out executable files to all kinds of 'peripheral device[s]' (UAVs, Uninhabited Ground Combat Vehicles [UGCVs], smart munitions, smart minefields, or smart, infinitely variable multipurpose 'blobs'). The soldierboy body, a steel rack covered in weapons and sensory devices, can be tossed about harmlessly at any speed or gravity: 'The bail [a grab bar] snatched us with an impulse shock of fifteen or twenty gees. Nothing to a soldierboy but, we found out later, it cracked four of the woman's ribs' (Haldeman *Peace* 79). Twenty gravities ('Gs' or 'gees'), twenty times the force pressing us down on the Earth's surface, is enough to kill or injure a human. But to the remote-brained soldierboy, no ribs crack because there are no ribs left.

Current, which means obsolete, military design proposes a lightly armoured soldier accompanied by 'remote sensing devices – not physically attached to the soldier. "Robotic mules, dogs and eagles can provide intelligence and reconnaissance, and can provide over-match effectiveness in collaborative real time planning and execution," Wallace said' (Book 'Competition' 33). BMIs will be essential to the robotic ark Wallace elaborates. The rest of the gray language, and particularly the word 'collaborative,' refers to the infamous 'system of systems' (networks of high technology that rely on other networks) at the heart of the Revolution in Military Affairs. The human has become a remote, distributed entity that will have to be woven into the system. In *The Forever War*, junior officers are fully immersed in a training simulator, a 'machine [that] kept my body totally inert and zapped my brain with four millennia's worth of military facts and theories. And I couldn't forget any of it! Not while I was in the tank' (174). Attuned to the gray obfuscation of military language, Haldeman calls the trainer the Accelerated Life Simulation Computer (ALSC). After three weeks, the subject crawls out fully loaded, all recorded human knowledge of battle, tactics, and methods of combat now burned into the mind. Forcing battle codes into the brain, like overtraining the body, was one early way behaviourists prepared the soldier to stay active during battle; it skirts the issue of battlefield shock and fear, although it in no way avoids the soldier's later struggle with war trauma. In *Forever Peace*, Haldeman reasons that his tele-commuting soldiers will suffer high stroke and heart-attack rates: soldierboy feedback systems make battle seem real enough for the body which, while it can no longer be torn by projectiles, can be attacked by its own core systems of the brain and heart.

Soldiers on the conservative side of the RMA worry that their futures may be nightmares of connectivity, with them becoming the 'blobs' to which DARPA attaches wires. Much combat training is already virtual, accomplished through the Close Combat Tactical Trainer program (CCTT). The CCTT is part of the Program Executive Office for Simulation, Training, and Instrumentation (PEO STRI) division, whose shoulder patch reads tellingly: 'All but war is simulation.' The motto fuses the actual and the false, viewing only war as reality. Connected training is essential for soldiers wearing helmets with high-speed interactive displays who face physical and information war: 'We are planning to place the ability to see the whole battlefield in the hands of every soldier in a way that has not been possible since the nineteenth century. This may open the door once more to shock' (Bateman 39). When Bateman says 'see' he doesn't mean to look out, but in. Soldiers will wear helmets (as most pilots do already) with mounted displays that beam tactical and geographical data right into their eyes. Looking up for such a soldier means to see into the squad and even platoon or battalion's information pool, to look up into the stove pipe or chimney of command. The nineteenth-century battlefield, because it was largely without aerial observation or telecommunications (there were exceptions), and because weapons had not yet emptied the field of men, was visible to most soldiers. Battles were fought at relatively close order, even though chaos was usually the centre of that order. Bateman and his fellow soldier-scholars worry that troops simply will not be able to process, or, to use a more organic term, soak up, the information shooting into their eyes and ears: the problem is caused by trying to 'access enough information simultaneously' (Bateman 17). Information overload and its resultant shock paralyses humans who are not subjected to potentially fatal situations: battle would exacerbate, rather than reduce, such shock.

In 1959 Robert Heinlein wrote off worries about the soldier's predicament on the high-speed battlefield with a single line: 'I've had the injections, of course, and hypnotic preparation, and it stands to reason that I can't really be afraid' (5). Being human, Heinlein's soldier *is* afraid, but that doesn't slow him down during combat. Heinlein's picture of an open battlefield where his soldiers bounce around in miraculous armour, dropping nuclear warheads as needed, doesn't match the complexity of contemporary war zones, particularly when the Army knows that most battles will be Military Operations in Urban Terrain (MOUT). The Army's first-and second-hand experiences in Mogadishu and the Israeli-occupied West Bank has reminded the post-Vietnam generation of soldiers that, if women and children are prepared to join in street fighting there will be no 'clean kills,' no innocents; war will be sudden, close, dirty, deadly, and total. In urban terrain where all are potential combatants, there is too much to see, too many decisions to make per second, and not enough brain time to discriminate between war and atrocity, which is difficult at the best of times. Strain on the soldier will likely appear in two very different ways: paralysis brought on by the inability to make informed decisions, or indiscriminate action where the soldier concludes that 'all before me are the enemy.' These poles are separated by a continuum of reactions; soldiers will rarely respond in only one way – but the damages will be high. Haldeman argues that even virtual soldiers pay hard physical costs: 'All ten of our soldierboys came into the garage within a couple of minutes. The mechanics jacked out and the exoskeleton shells eased open. Scoville's people climbed out like little old men and women, even though their bodies had been exercised constantly and adjusted for fatigue poisons' (*Peace* 4). While the mechanics extricate themselves in the midst of violent images of plugs being pulled, sensation ending abruptly, the machinery is at peace, easing open. Armour has banished all flesh but brain tissue, and still the strain has withered the humans into ancient people. If simulation makes the body believe it's real, then stress in simulation is real stress; wounding is wounding; atrocity is atrocity; death is death.

Joe Haldeman arrived in Vietnam on 29 February 1968. Army personnel rotated home a year to the day after arriving in country. In his nightmares Haldeman expected to be caught in Vietnam forever. Time, its nature and cessation, its eccentric collapse and expansion caused by pain, anxiety, grief, is Haldeman's metaphor for war's unshakable trauma. In *The Forever War* Haldeman's final time device is a stasis field in which all electrical activity ceases and body armour continues to function only because it has been specially coated. Since electrical activity stops, energy weapons no longer function. After a millennium of fighting with advanced arms, the new starship troopers are outfitted with bows, darts, quarterstaffs. Haldeman's unblinking narrator is reduced to this: 'I drew my sword and waited' (*War* 239). Haldeman seems to argue that the logic of war technology is circular: in *The Forever War* the soldiers return to the weapons and tactics of the phalanx; in *Forever Peace*, the enemy pounds on the soldierboy with rocks. Just as the machine-gun held the world steady in its sights for four murderous years, armour promises a battlefield that paradoxically lives on the speed of dataflow and overload, and is static in time where phalanx meets phalanx and paralysis results. If we are uncertain about the kind of stasis Haldeman depicts, we should remember that, for decades, we have lived in a similar field wired together by the twisted logic of nuclear Mutually Assured Destruction (MAD).

Armour will dominate the battlefield, commandeering the resources of science and engineering agencies dedicated to technowar – the same agencies will likely be hired to determine how to incapacitate armour. But, while the logic of armour looks static, or even reductive where we return to swords and rocks, we can see that it is progressive. Armour has rested on, penetrated, slid into, then abolished skin. Armour is breathing in humans and breathing out flesh. Robots guided by remote operators wearing Brain Machine Interfaces will fight each other over vast expanses of space now rendered entirely toxic for human habitation. As the operators break down or have aneurisms, heart attacks, and strokes, they will be replaced by younger people with better reflexes. And armour will progress to its next evolutionary position.

Chapter Three

Heavy Tread: On Track for Battle

The resolute imperial will was all played out here at the empire's fringe, lost in rancor and mud. Here were pharaoh's chariots engulfed; his horsemen confused; and all his magnificence dismayed.

A shithole.

– Tobias Wolff, *In Pharaoh's Army* (1994)

The main battle tank is a new kind of body, a seventy-tonne low-slung whining turbine-driven 1500-horsepower gyro-stabilized fire-on-the-move slick war-utility vehicle. It's grown fat in recent years, now carrying almost a metre of rolled homogeneous armour that is smart and reactive. Armour plate, tank muscle, is calculated first in tens, then hundreds of millimetres; strength is a physics of ballistics, muzzle velocities, gun-tube lengths. The tank story is about a different body, one that rolls away from human scale. Human arms throwing missiles have been replaced by 120 mm smooth-bore gun tubes that fling javelins of depleted uranium at five times the speed of sound, heavy poison darts that fly through solid ground, then one or even two tanks at a time, before their kinetic energy dies off. Twenty-kilogram shells the size of a human thigh arc downrange four to five kilometres where their targets aren't hit so much as obliterated. Tanks that have been crippled, or don't have enough armament or armour plate to compete in battlespace, explode like popcorn when attacked. The centre of the vehicle vanishes in a greasy metallic sludge that is rendered down to its basic minerals and fats. The tank story is about time travel; tank and anti-tank weapon are locked in a spiral of mechanical evolution. A tank from the past meeting a tank-killing weapon, like a rocket, from its future will perish: anti-tank weapons that meet new armour are fatal to the infantry. Inside and out, it's a hard story.

Tanks, originally a secret weapon, were first used by the British in the First World War to help the infantry move across no man's land.[11] They were slow-moving (five kilometres an hour), heavy (seventy tonnes or more) steel shields designed to break the machine-gun's grip on the battlefield. The famous 1917 battle of Cambrai, in which tanks rapidly took such a mass of ground that the infantry couldn't hold it, demonstrated to the British and German high commands that industrial warfare had altered again; the stasis typifying the First World War could be eliminated by the tank's relative speed and mobility. British tank theorists Ernest Swinton, J.C. Fuller, and Liddell Hart understood the implications of the tank for the future of warfare. First World War German staff officer Heinz Guderian saw what Swinton, Fuller, and Hart did, but there was a signal difference between them: despite everything it had learned from the First World War about tanks, the British high command dismissed Swinton, Fuller, Hart, tank theory, and tanks altogether, while Hitler embraced them – in April 1939 Fuller visited Berlin and, having been shown mobile armoured forces, was asked by Hitler: 'I hope you were pleased with your children?': Fuller was very pleased. Guderian elaborated on the other men's work, formulating the concept of *schwerpunkt* (emphasis or focal point) in which the tank brigade became the centre not only of the battlefield but of a whole campaign. Guderian taught tank theory to a generation of Second World War tank commanders, watched tanks in action in the Spanish Civil War, and in *Achtung-Panzer!* (*panzer* means armour), his 1937 meditation on the coming of tank warfare, implicitly promised an unnamed commander success if he followed Guderian's advice: 'Will we try to resolve an imposed deadlock by one mighty, concentrated commitment of our main offensive weapons? Or will we renounce their inherent potential for speedy, far-reaching movement, for the sake of tying ourselves to the snail's pace of the infantry and the artillery batteries, thereby renouncing all prospect of a speedy decision to the battle and the war?' (168).

Guderian's questions hammer at the German army's unease with a heavily mechanized, far-flung industrial force: he attacks its preference for light infantry tied to horses, short supply lines, and low technology. He argues for a shift into the mechanical future where tanks, once burned up in support roles, will govern the field. His dismay at the scattering of tanks in what he calls 'penny packets' across the battlefield creates the military argument for deep and broad attacks into enemy lines. Tanks must carry humans, humans must come up to military speed. In Guderian's accounting, mechanized warfare would lift humans out of the trenches (all the early tank theorists were similarly impassioned, arguably messianic, about the technology), would carry other recent tools of science's gifts to war (poison gas, flame-throwers, machine-guns, and especially high-explosive, time-delayed artillery shells) forward into enemy territory. Movement in rapid land warfare will be prepared for not by artillery barrages but by low-flying air support: this is what newspapers, not the military, called *blitzkrieg*, lightning war. Combatants and civilians currently trying to survive AirLand Battle as practised by the United States recognize Guderian's philosophy. The tank suggested a new way to keep the soldier body safe: rather than cowering in trenches for protection, the soldier could wear a steel coat. What tankers didn't see coming was the personal body armour that would arrive in the next century, nor did they understand how tight human connections to tank bodies would become.

The pre–Second World War British high command was satisfied that tanks were to act as infantry support, to cut wedges across entrenchments. Before the Second World War started, both the British and the French soldered themselves to their technological mistakes: the British wrote off tanks as offensive weapons, the French inscribed their useless Maginot line on the land. Both countries were still chained to stasis when Guderian's doctrine rolled over them; both looked to frozen or slow-moving low technology to handle the new industrial onslaught. Coming wars were not to be so much between infantry and infantry as between rapidly evolving weapons systems. Tanks didn't aim at men, as the British expected, but at each other: 'War showed that tanks shot at tanks more often than they supported infantry, and the dedicated role of the tank destroyer failed to materialize' (Hogg 113). Fields in the Soviet Union and deserts in North Africa filled unexpectedly with tank battles. The Soviets adopted J. Walter Christie's advanced tank-suspension systems even while the British held on to long-extinct First World War tank designs.

The famous Soviet T-34 (even the Germans called it the best tank of the war) illustrated the usefulness of rounded or sloped front plates. The tank's front face, called the glacis (a term borrowed from siege fortification that indicated how much tanks were originally seen as moving fortresses), is the most obvious and attractive target for enemy fire. Between the First World War and the present, the glacis on all main battle tanks (American, Israeli, British) has reclined steadily from an angle near 0° to 80° (Clancy 7). The more upright armoured plate, the less material a shell must penetrate: a sloping glacis means more armour (German anti-tank crews discovered the impact of the new geometry when their armour-piercing shells literally bounced off the rounded sloped glacis of Soviet T-34s). At the same time, the glacis has been thickening up (armour is often referred to as RHA, Rolled Homogeneous Armour, which until recently has proved to be the toughest plate available). A thick, sloped glacis means that a direct hit now has to plow through up to a metre of armour. As with ships and air- and rotorcraft, utility has pushed machines into flatter, harder, sleeker shapes that Guderian would call 'more technically "beautiful"' (137). Where once whole tank schools argued about whether a tank should stop and fire or fire on the move, now all main battle tanks, the American M1A2 Abrams, the Israeli Merkava Mark 4, the British Centurion 105, roll forward at seventy-two kilometres per hour, guns traversing sinuously and firing accurately, independent suspension and stabilization systems making the turrets look like entities separate from the chassis. Tanks are increasingly invisible to radar, have stealth profiles, and use wide treads to spread their considerable weight over a lot of ground. Never have tanks been simultaneously heavier, faster, more powerful, safer for the operators, and more dangerous to opponents. Despite their weight, which makes them unappealing to some RMA adherents, tanks are still central to the war-fighting doctrine called AirLand Battle. AirLand models emphasize small-unit manoeuvre warfare instead of the massed armies that typified the Second World War and Warsaw Pact doctrine. One of the more recent arguments for AirLand Battle is Harlan Ullman and James Wade's *Shock and Awe* (1996), where deep, fast, synchronized attacks that, through manoeuvre, agility, small-unit initiative, and shock to the enemy, bring about an abrupt end to combat (often the attention is given to the 'first day of war' in which the whole war can apparently be decided). AirLand Battle, as the fused words indicate, aims to eliminate wars of attrition like Vietnam by relying on light rather than heavy weapons (the recent move towards the Stryker light armoured personnel carrier and away from the M1A2 Abrams tank is one such example). But tanks have proved themselves necessary even in the twenty-first century, and they have consumed much human blood as well oil to bring them this far in their evolution.

The story of tank design can be organized by two geometric shapes: a triangle, and a circle. The triangle of mobility (speed), weight (armour), and firepower (tank cannon plus ammunition) forms the concept of a tank. Somewhere in the triangle is a balance point between the three. To make a tank fast means surrendering weight in armour and firepower; a well-protected, heavily armoured tank makes for sluggish movement, one that can wiped out by more agile weapons. A tank that carries a heavy gun must also carry heavy rounds for the gun. Without speed, an immobile juggernaut will eventually be destroyed. Without armour, a speedy machine will be destroyed. Without sufficient firepower, even a tank that can fire and move quickly will be destroyed by a tank with a long-range gun. Properly drawing the triangle of speed, armour, and firepower will result in what the U.S. Army calls 'survivability' – both the weapon and its operators will survive encounters with the enemy. The triangle must be reorganized and redrawn each time tanks are designed. In the Second World War the German *Panzerkampfwagen* (PzKw) VI Tiger tank was superior to other tanks in armour and gun range, but the best overall design triangle was represented by the Soviet T-34.

The second geometry, one that encloses the first, is the circle of the clock-face. If we consider 'tank' as an abstract, an armoured, mobile gun triangle drawn differently by each designer, then when the triangles meet in battle it will in large part be the luck of the design draw which is most successful. A ponderous tank with greater reach and protection (heavy on the gun and armour side) will succeed where a lightly armoured fast tank, a speed isosceles, cannot. Because the triangles take a long time to design and manufacture in any quantity – Guderian's vision of large collective tank action was proven luridly by Germany's conquest of Poland and France – and retooling cannot take place immediately, then once a campaign has started there isn't a lot of time for adjustment. Soldiers fight with what they brought. Battlefield modifications to equipment, while they seem sensible to the operator, will not always increase the soldier's chances of survival. New technology is not necessarily better if weapon design is poor: reliability and serviceability are also players in technological struggles. Smart weapons' designers look ahead to imagine what their weapons *might* need, and then consider and reduce the weaknesses of their current creations. New design emerges from what should be, not what was. If one weapon is even slightly ahead of another and is present in sufficient quantity (only short material supply and lack of field support prevented the PzKw VI 'King' Tiger from ruling the Allied battlefields), then the prospect for the technologically duller side is grim. Design time governs all else. The clock encircles the triangle: weapons from the future will demolish the enemy.[12]

Here are two geometries on the same plane: the first is the relatively fast, lightly gunned (75 mm cannon) and armoured (various thicknesses of armour, but usually under 100 mm) American M4 Sherman. It was the main battle tank for the American (and many British) forces as they advanced across the Normandy countryside in 1944. The M4 Sherman came up against the much slower, more heavily gunned (88 mm) and armoured (100 mm of RHA) PzKw V Panther and PzKw VI Tiger tanks. Tank historian Ian Hogg explains the differences: 'In real terms it meant that the Tiger could knock out a Sherman at a range of 3,500 m (3,825 yd.) while the Sherman had to get within 200 m (220 yd.) to do any damage to the Tiger with its 76.2 mm (3 in.) gun, and with a 75 mm (2.95 in.) gun would probably do no damage if he laid alongside the Tiger and fired'

(159). Tigers withstanding hit after direct hit, destroying strings of Shermans, are, for the Allies, disastrously familiar battlefield geometries. The Shermans prevailed, as Belton Cooper explains in his aptly named memoir *Death Traps*, because American production lines turned them out prodigiously, men kept stepping into them, and the Germans ran out of materiel and soldiers.[13] Tank discussions inevitably come down to numbers: Cooper can be specific because he had to calculate the daily loss rate of tanks in his sector so that his unit could send for replacement vehicles (not crews; which, after a few months, were assigned from untrained soldiers with no experience in mechanized, or armoured, battalions). The staggering numbers of Shermans and their crews incinerated daily drove Cooper into a rage not only with the army but with General George S. Patton specifically, who had chosen the M4 over its heavier rival the M6 (fielded in 1945 as the M26). Cooper's rhetoric is full of loops and repetitions, and his narrative at times stutters, caught up in surreal eddies familiar to Joseph Heller's characters in *Catch-22*. The weak Shermans were refugees from an earlier, more naive design cycle when weapon makers couldn't, or wouldn't, imagine tanks simultaneously as heavy and mobile as the Panthers and Tigers.

Soldiers on the wrong end of design geometry find that their 'will to survive increased the innovative spirit,' prompting many illegal battlefield modifications (Cooper 90).[14] In both Normandy and the Pacific, tank crews begin defending themselves and their machines against magnetic mines by bolting thick oak planks, or even pouring concrete, over the armour; mines now exploded on their surprised users. Mines or satchel charges thrown on relatively lightly armoured vehicle surfaces were defeated by additions of steel netting or fencing bolted a half-metre or more above the turret; the airspace between the mine and tank reduced the blast enough to allow the tank's armour to deflect the rest. Even in the higher tech days of Vietnam, tankers took to adding 'perforated runway matting ... one sheet to each side, we hung extra tread sections around the turrets, stacked sandbags around the command cupolas, and filled the bustle racks [outer rear storage compartments] with ammo boxes and cartons of canned rations' in order to defeat rocket fire (Birdwell and Nolan 119). Instinctively, the tankers began to put armour on their armour. It hardly matters what the material is, nor does it matter that the sandbags, cement, and runway matting unbalance the machines and put pressure on axles, torsion bars, and drive shafts that were never rated for it. Whether a sandbag (or can of beans) will stop a rocket round or not isn't the issue: the soldier is on the wrong side of the protection triangle, knows it, and seeks to strengthen the outer body's armour. Strengthening armour is not so much about weight as about philosophy.

Famous Israeli tank commander and designer Israel Tal not only redefined what a tank is (a body of bodies, where the crew must be preserved in a mechanical placenta) but also developed Explosive Reactive Armour (ERA). When a shell hits ERA, the armour responds with its own outward blast that deflects the attack. Cultural historian Patrick Wright notes: 'As the philosophical Tal says of this ingenious dialectical manoeuvre, "We make the projectile commit suicide. The energy of the projectile is used to defeat it"' (325). For a tank to get stronger, it can't simply pack on more weight but must teach attackers Sun Tzu's lessons in the art of war: the enemy's strength is a defence. Missiles will blow *themselves* up. ERA was part of the solution tankers in Normandy and Vietnam were looking for; the rest of that solution has come from what is known as Chobham Armour (named for the lab that developed it). Chobham Armour is a combination of 'steel armour into which is embedded such things as tungsten rods and blocks, interlayered with plastic and ceramic materials' (Hogg 30). Always meticulous with detail, Hogg is forced to use the vague phrase 'such things' because Chobham Armour is a classified product about which tank historians know only a few basics. The combination of materials (including airspace) is what makes the armour both light and powerful – it isn't all steel plate. When a shaped warhead (most anti-tank missiles are shaped charges) explodes against Chobham Armour, the intense heat and pressure renders the ceramics into a sort of fluid through and around which hot gases flow. The space between the blocks diffuses the pressure waves produced by the shaped charge into smaller, less harmful superheated jets. The steel layer prevents the long rod in the warhead from penetrating the tank hull. Were a projectile to shatter the inmost armour, flakes of the tank's own metal would fill the tank in a process called 'spalling.' Anti-spalling liners have been added to the most recent

American M1A2 Abrams tanks along with, it is rumoured, a depleted uranium mesh that has been incorporated in the hull. All main battle tanks are equipped with Chobham Armour, anti-spalling liners, and cabin over-pressure, the last offered as protection against nuclear, biological and chemical (NBC) weapons (over-pressure means that airflow into the crew compartment keeps the air at a high enough pressure such that NBC agents are forced out of the turret).

It may begin to sound as though the interior of a tank is a reasonable place to be, so it's worth considering the main tank-killing instrument, called an Armor Piercing, Fin Stabilized, Discarding Sabot (APFSDS) round (fig. 7). The APFSDS round, usually known simply as a 'sabot' because of the 'shoes' that fall away once the missile has been fired, has at its centre a long thin depleted uranium (DU) dart. Because the dart is much thinner than the tank cannon's 120 mm bore, it is encased in a metal ring that centres it in the barrel. The ring, formed of three pieces of metal, reach out to the barrel and hold the round in place as it shoots down the unrifled tube.

DU has replaced tungsten as the sabot designer's core metal of choice because it is harder, cheaper, and more plentiful than tungsten. The fins are necessary to stabilize the round in flight (so, Fin Stabilized). Once out of the tube, the sabot discards its shoes (the remainder of the acronym: Discarding Sabot), and, at about 1,800 metres per second (mach 5.3 – over five times the speed of sound, a speed possible only because there is no rifling in the gun tube), flies at its target.

When the depleted uranium dart pierces armour, it enters the tank as a rainburst of super-heated uranium splinters that shower the crew compartment, cutting or burning everything they touch; the rest of the DU becomes poisonous radioactive vapour. The sabot is not an explosive round, but, because it's made of a pyrophore, it acts like one. Even against Chobham Armour the sabot is deadly. In the first Gulf War, M1A1 Abrams tanks fired sabots that flew through hardened berms and *then* penetrated heavy Russian T-72 tanks. Sometimes sabots blew through two tanks in a row. In one of the U.S. friendly fire cases, M1A1 tanks fired on American Bradley fighting vehicles, a lightly armoured troop carrier, destroying the vehicles and killing their crews – one Bradley was deemed too radioactively hot to be moved and was

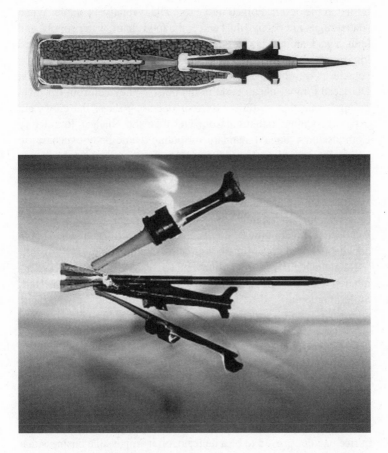

7 **The Whole, Apart**: A cutaway view of an intact APFSDS 'sabot' round (top) looks like a stubby mechanical pencil, where the tip of the depleted uranium rod would be the pencil point. The dark collar that grips the round's point comprises three pieces that make up the shoes. Inside the shell casing, at the left end of the dart's narrowing cone, is a set of fins that act like an arrow's fletching. The rest of the shell is packed with 'propellant' (explosive). When the round leaves the gun tube, it splits open to reveal the 'long rod penetrator' DU dart, which, in the lower image, drops its shoes.

buried in the desert. Forged into a rod, DU is relatively stable. When shattered, its radioactive dust enters the food chain, water supply, and human and animal bodies, with predictably grim oncological and long-term genetic results. There is not much 'survivability' for a tank, vehicle, or crew hit by a sabot round. Tank crews also have to hope that DU-laced armour doesn't add to the poisons they confront (medical groups argue that, within sixty-four operating days, the crews are exposed to more radiation from DU than the Nuclear Regulatory Commission's 'annual standard for public whole-body exposure' to radioactive materials (Fahey 29).[15] The unsteady movement between tank and anti-tank gun, the evolution of better armour, then more sophisticated missiles with which to pierce it, is what is called the 'gun-armour spiral,' an apparently endless cycling between defence (heavier, tougher, armour) and attack (deadlier tank-killing weapons). But the question of how to step out of the loop is difficult when being outside, exposed to automatic weapons, is unthinkable.

Being inside is a story of allure and terror. Mobile, fast, tracked armoured gun wagons appealed to their originators (Swinton, Hart, Fuller, Guderian), all trench war survivors. But soldiers who came after them with no experiences of trench stasis are more ambivalent about tanks, which they see as domestic spaces that unpredictably draw fire, catch on fire, and kill their residents. Mechanized soldiers in Vietnam, whether in one of the six hundred tanks or other armoured vehicles (particularly the M113 Armoured Personnel Carrier [APC, known colloquially as a 'track']), found themselves in strangled relations with their machines. The Vietnamese made an art out of the daily mining of roads and jungle trails; using heavy armour in rice paddies proved to be a delicate, often impossible business that gave the word 'quagmire' a certain literal quality. Larry Heinemann's 1967 *roman-à-clef Close Quarters* about a mechanized unit witnesses the soldier's frustration. The protagonist arrives in camp to see one exhausted veteran emerge from an APC:

> He brought the spade back over his head and began beating on the front of the track with the blade end, like it was an axe.
>
> 'You goddamned half-stepping fuck-up track!' He said it low and slow and mean. 'I'm gonna tear you apart with my bare fucken hands.

I'm gonna pour mo-gas on your asshole and burn you down to fucken nickels, you short-time pissant mo-gas-guzzling motherfucker!' (Heinemann 6)

For a year the soldier has borne the fear of riding in or on a steel coffin: when the machine is 'buttoned up' (completely closed), it will become an instant charnel house if hit by a rocket or mine. If open, the APC gives the soldier high ground from which he can see, and also makes the soldier a sniper's gift. Heinemann captures the crazed frustration of being dually helpless; the soldier comes at the APC first with a shovel, then with his hands, finally with the deepest threat he can summon: complete dissolution at the foundry – only industrial destruction can unmake the APC and reduce it to harmless coins. The APCs are seen by the protagonist as great snorting beasts; this soldier also finds body parts in the machine, reflecting the familiarity between the soldiers and armour as well as the collapsing line between what is alive and what is not. APCs built up a long and difficult history in Vietnam, starting with their first, famous disaster in 1963 at the battle of Ap Bac, where a squad of inexperienced but trained, terrified North Vietnamese soldiers pinned down or destroyed a number of new APCs (Sheehan 203ff.). On paper, APCs looked good: lightly armoured, relatively heavily gunned, tracked not wheeled, easily serviced, able to run on gasoline, they seemed to be both the logistical and the tactical solution to protected troop movement (when they failed their initial promise the helicopter would be seen as the next magic bullet). The dread of ambush and mines (the North Vietnamese became famous for remining roads that had been swept the day before – nothing was as it seemed, and nothing was safe for Americans in Vietnam) kept the troops riding in sandbag bunkers on top of the tracks: 'I know, I've seen,' says Heinemann's protagonist, 'the tracks burn like paper soaked with tar ... and the drivers, the Deltas, almost never make it out' (102).

And yet M113 survivors write nostalgically about their machines. Many of the crews experienced what most tankers do: the sense that armour, despite its cost, provides the only protection there is. Rather than camp under a poncho (many tankers look at the infantry with sympathy, unable to believe how much they must and do carry), APC crews use the vehicles as bunks, hot-rods, mess halls, sex parlours, and, above all, reassurance engines. Soldiers who had survived far into their tours refused to leave the machines in which they had stored everything emotional and physical remaining to them. The APCs were stable, portable elements of American life, battle double-wide trailers in the deadly Vietnam camping ground. Almost all tankers become make-shift mechanics complete with the love of engines that many young men display. M1A1 tank commanders in the first Gulf War learn their way around the big tanks, 'becoming familiar with its nooks and crannies and secrets – to return to the marriage trope, taking our first nervous gropes toward learning to be intimate with her. When I left the army two years later, I had domesticated my perception of the tank: I had grown to marvel over its compactness. So much power, so many memories, bundled in such a svelte package' (Vernon 45). Vernon isn't the first soldier to describe his tank in terms of sexuality, domestication, and the reassurance that long-term relationships bring. Adolescent fumbling encounters becomes suave sexual certainties when the soldier kills, or scores. The tank is a hard body into which soft flesh slides: when the hardened womb saves the crew's lives, it becomes a miraculous device. The tanker's body sloughs off – it is only real when inside armour. Foreplay in nooks and crannies gives way to the deep thrusts of long rod penetrators: 'I did not get the rush from firing the 25mm [heavy machine-gun] like I did when I fired the M1's 105mm main gun' (Vernon 103). Nooks and crannies, firing slits, the turret's swivelling hips, all blend together when the boy becomes a man: 'When an M1 fires, the earth quakes.'

A rolling gun that shakes the earth is visceral proof that the enemy can be stopped. The terrifyingly fast German advance into Russia was finally met by the stolid T-34; a great deal of solace could be drawn from the T-34's ability to survive – Russia could see itself in a tank so robust that it 'remained undamaged in spite of the fact that, as was later determined, it got fourteen direct hits. These merely produced blue spots on its armor' (Perrett 78). The T-34 couldn't stand up to the longer-range, higher-powered guns of the Panthers or Tigers, but, as with the American production of the Sherman, the sheer number of T-34s on the field guaranteed their eventual success.[16] The 'blue spots' were like the welts raised on Russia's body by the Wehrmacht. Only armour could stop the German advance at Kursk in 1943 where some six thousand tanks fought in the world's largest tank battle: armour must meet armour if the infantry is to survive. Soldiers facing armour (Germans against Russian T-34s in 1942, Americans facing the Panthers or Tigers in Normandy, Japanese fighting American M3 Grants on Pacific Islands, Iraqis confronted by M1A1 Abrams [1991] or M1A2 Abrams [2003]) know that being on the wrong side of the design wedge is fatal. Whether superior tanks incorporate one or all of better suspension, armour, guns, engines, or treads, all soldiers recognize a winner where battles become slaughters. Belton Cooper recalls one engagement where the American forces 'started with sixty-four medium tanks, and we lost forty-eight of them in twenty-six minutes. A proportional number of soldiers died in this terrible fight' (148). The range of the German guns, the lightness of American armour, and the scarcity of veteran American tankers put the Americans in a hopeless situation. This was not the war of attrition that Guderian and Fuller so much feared – instead it was a twenty-six-minute turkey-shoot. Fifty years later, an Iraqi battalion commander recounted: 'When I went into Kuwait I had thirty-nine tanks ... After six weeks of air bombardment, I had thirty-two left. After twenty minutes in action against the M1s, I had none' (Clancy 62). Aerial bombardment cannot compete, according to this story, with ground forces that arrive from the future. In science fiction author Leo Frankowski's vision, the time taken to decimate the enemy will drop to seven seconds (96). The future arrives in the past with a great rush and slaughter unites them.

The tank limns the increasing fragility of the soldier's body. Absorbed by the army and digested by war, the body seems to have disappeared in bureaucracy and mechanism. Living directly in the machine's belly (or womb, depending on how one understands the tank's polymorphism, with its permanent erection connected to a steel life-support enclosure) has great benefits for humans facing artillery, air assault, and machine-gun fire.[17] The tank also vividly illustrates the commonness, in both the crude and shared senses, of the soldier's body. If one tanker suffers an indignity, all will: 'Like going to the bathroom,' says one soldier. 'It's amazing how people never think about that. If you got out you got killed, and if you stayed in there you shit your pants.' It's a reasonable, if unpleasant, reality, but it's not the end of the narrative: 'It wasn't near as big a problem as it would have been. Unless you had diarrhea' (Gilbert 27). Because this soldier fought in the Pacific, the likelihood that he and the rest of the crew *did* have diarrhea changes the situation. It wasn't just a matter of shitting once but of doing so repeatedly, helplessly, painfully, publicly. The tank is transformed into a large sewer, permanently filthy, wet, and smelly. In this same tank the crew must eat and, often, sleep. The only time tanks were washed was when their crews were killed and the vehicles were being refitted (in the Pacific war there simply wasn't enough water, if the troops got inland, to wash anything). The body is suddenly captive, and things preserved as exterior and private by industrial civilization's amenities (the body's waste-disposal system externalized by the flush-toilet or even the 'privy'), are now internal and dreadfully public. Bodies and their indignities, their shames, are seen, heard, felt, smelt. There will need to be new rules about what is offensive, shameful, visible, if the men are to continue living with themselves. A hard-charging tank sergeant in Vietnam gets an unexpected body lesson about armour when his M48 Patton hits a mine. Sore all over and almost unable to move (which for this man, even when wounded, is a rare event), he is told by the unit's surgeon: 'Basically there isn't anything I can do. You're just loosened up inside – the ligaments that hold your organs in place are weakened' (Zumbro 166). Bodies weren't designed to be put into metal boxes and rolled over high explosives; the threads that bind bones together can now be felt only too well, the body's fabric has been uncomfortably stretched.

But these are just indignities. The soldiers who stay inside despite the shit, the shame, or the loosening of the body's internal harness, and survive, are one group. Rage and horror are the portion left to tank-recovery crews. Forced to clean up after literally hundreds of slaughtered men, Belton Cooper learns that when 'a tanker inside a tank received the full effect of a penetration [a shell that explodes inside the tank], sometimes the body, particularly the head, exploded and scattered blood, gore and brains throughout the entire compartment' (20). Within a few days of his arrival in France, Cooper must clean out a tank, so he's used to the disaster inside. It's the identifiable pieces that make the job almost impossible. Late in the Normandy campaign, when the soldiers should all be deeply psychically numb, they are still unable to face a recovered tank where a German long shot (literally) has blown down an M4 gun tube and exploded through the open breech. The clean-up crew, in fact the whole unit, is unable to face removing and separating the two victims' body parts. Finally, an unassuming volunteer steps forward and the recovery begins (Cooper, modest about his own daily heroism, explains in detail why this volunteer is a true hero: he can face the body's utter ruination and still see the corpses as human beings). The tankers know how much of themselves they will leave behind: unless the tank burns completely (a fairly common event called, in that particularly teatime cheery way, 'brewing up,' by the British), *everything* will remain. An unrecognizable vapourized spattered everything will be scraped off the walls, the tanks will be repainted with lead white, renumbered, and sent to different units where new crews don't know the tank's history.

One tanker estimates his chances of getting out the tank's emergency trap in the event of a hit:

> At best you would have ninety seconds to get out that door; if it jammed you would need fifty of those seconds to push it open. That would leave forty seconds for three men to squeeze out. Tick, tick, tick, boom? And what would happen if both the turret and the trapdoor were inoperative? What would happen is, you'd die! It takes twenty minutes for a medium tank to incinerate; and the flames burn slowly, so figure it takes ten minutes for a hearty man within to perish. You wouldn't even be able to struggle for chances are both exits would be sheeted with flame and smoke. You would sit, read *Good Housekeeping* and die like a dog. (Wright 154)

Gallows irony thickens the reality of roasting to death in a tank. The soldier's calculation is thorough, complete with body weight, vehicle type, the specifics of extermination. His final act will be to raise a metaphorical middle finger to the bourgeoisie, the keepers of houses and gardens he won't ever visit, those for whom he has been told too often he is 'making a sacrifice.' The tanker can consider himself lucky if his burning tank is full of ammunition: the fire will be explosive and death quicker. Unlike field-artillery units that establish separate ordinance stores, tanks are necessarily moving ammunition dumps carrying their rounds inside the crew compartment. The terror of a fully loaded shell rack 'cooking off' (where the heat inside the tank ignites high explosive in stored rounds) became so great that the Abrams' designers installed blast-proof doors which, when shut by the loader, would direct explosions from the tank's ammunition rack outward. Before then, tankers lived unprotected in rolling magazines. The crew that survives its first battle, having witnessed the manner in which tanks are destroyed, becomes quite grim about its future (and the probability of having one). Despite different nationalities and cultures, almost all tank accounts display the mordant humour of those who have already forfeited their lives. But that wit can't divert tank crews from the picture of the machine that burns completely, where the 'trapped crew had been broiled in such a way that a puddle of fat had spread from under the tank and this was quilted with brilliant flies of all descriptions and colours' (Wright 154). Metal hits metal, shell on tank, and, as the metal is smelted, the flesh is rendered too. Flesh pays the price, and the bodies reappear in a horrid form, like the brains that Cooper saw spattered in tanks, even more visible for their alteration. If the crew's bodies simply vanished, as occurs so often in artillery attacks on infantry, perhaps each soldier would be forced to conjure up his own horror of what had happened. But a puddle of spreading domesticated fat that has become a bedspread is impossible to ignore, guaranteeing the story a long, vivid life.

Flesh's fate under metal increases the soldiers' terror in Normandy as they climb back into weak M4 Shermans and face far superior German tanks. Both armies know how short existence is for the crews of machines called 'Tommy cookers' by Germans (no differentiation between the British or Americans being roasted) and 'Ronson burners' by the Allies 'because of that firm's claim that their products [Ronson made and makes cigarette lighters] always "light the first time"' (Ellis 154). One strike by a Panther or King Tiger and the Sherman would become a pot of human stew (a similar domestic familiarity used to naturalize dreadful events was common to the eastern front, where German soldiers, watching the Tiger's powerful 88-mm. gun rip turrets off T-34s, said 'The T-34 raises its hat when-

ever it meets a Tiger' [Perrett 103]).[18] Tanks stop being human carriers and become lighters and matches at the hearth. The combination
of flesh and metal confronts Larry Heinemann's protagonist in Vietnam when he sees three burned armoured tracks that 'wallowed in
puddles of shining aluminum-alloy armor plate. Whole pieces were
scattered between, rounded and warped by the heat and soft-looking.
The gun-mount and decks, all the sides – everything but the steel
plates in the floors – had been incinerated. The rubber parts on the
tread cleats and solid rubber tires on the road wheels were burned
away, and the treads lay slack, like huge rubber bands ... And then I
saw the ridge of a temple, an eyehole, the bones of fingers at an arm's
length ... A body covered with that rubber and cloth smudge' (244).
The tank sergeant whose body was 'loosened' can now be paired
with the machines that have been stretched out of shape. At the end
of the tank's road is the complete elimination of its meaning: protection, force, mobility have become a pool. Only a slippery pliability
remains. The account begins in amazement at the magic shining
metal but soon seeks the bodies that are just as scattered as the road
wheels that drive the treads; the bodies are hidden in plain sight. The
skeletal remains of both machine and organic bodies are unified in
the morass they have become – indelible, indiscriminate, filthy
smears. At the end of the road is a war tar pit.

The infantry often expresses jealousy about armour but, once
inside, record feeling imprisoned, claustrophobic, blinded by metal.
The tanker, unlike the infantry soldier, has two bodies, two selves.
One steps inside another and may or may not emerge at the end of the
day's action. After enough time, the small human body becomes
accustomed to the large metal one – the attributes of dependency are
there: affection, interrupted by spasms of sheer hatred, for the enclosing other. The infantry hates tanks, but hates being inside them more.
The tanker hates tanks, but hates being outside them the most. The
larger body becomes the battle self, an early, rough version of powered armour. The machine extends the body and the self but at the
cost of panic when the tank malfunctions. Two tank commanders
from Vietnam, Ralph Zumbro and Dwight Birdwell, prove themselves spectacularly capable in the face of machine disaster. Zumbro
describes using water to slosh a fire out of his M48's full fuel tanks,

while Birdwell faces the disaster of not one, but two, rounds that jam in his M48's overheated gun tube. Fuel-tank fires and dud rounds in a hot cannon are both at the top of a list of undesirable armour events. Having yanked the lanyard and triggered the dud, now wedged in the barrel, Birdwell must determine how to proceed: Open the breech and risk the shell suddenly exploding into the turret? Or leave the round in, with its own dangers? 'The longer I ignored it, the more heat [the jammed round] absorbed – and the better the chance it might cook off as I removed it. I told Hines to vacate the turret for his own safety, then opened the breechblock, working the lever several times to extract the stuck shell. I picked it up, ignoring the pain – it was hot, hot, hot – as I frantically clambered up through the commander's cupola, then keeping low, prepared to throw it into the wire behind my tank' (Birdwell and Nolan 60). The minute he disposes of the one round, another jams: remembering to put on his asbestos gloves this time, Birdwell unhesitatingly ejects and dumps the second shell. Despite what they have seen happen to other crews, both Birdwell and Zumbro refuse to leave their tanks. Dedicated to the second body, tankers balance on a precarious, explosive edge. If they are to be able to proceed, they must imagine their bodies possessed of an unusual agency, that what happens to others will not happen to them: their tanks will not become Ronson lighters; fuel and ammunition will not explode while they stand in the hatch; they will not slip with hot dud shells. Their luck, but more, their ability, will hold. For the tanker who believes that it is better to be inside, external bodies, soft unarmoured bodies, will be conceived of as the foreign. The interior, familiar, comfortable part must be preserved; the exterior, outer, alien will be eliminated, sometimes in a frenzy of loathing.

During four years of trench warfare it was not uncommon for bodies of the dead, in various moments of decay, to reveal themselves as parts of the trench wall. As twentieth-century mechanized warfare rolled on, bodies, while they would still be used as fortifications, also became engineering materials: 'Driving over bodies became part of the degenerate routine of the Eastern Front. In October 1941 dead Russians were being used to further the German advance, their corpses being laid out as logs or planks in front of Germany's motorized units to assist their passage through marshy areas' (Wright

299).[19] From inside the tank we've seen one kind of body; outside, it's a story of difference. While it's grim to picture the bodies of the dead laid out before armour, at least they *are* dead. Long before the 2,000 kg Guided Bomb Unit 28 (GBU-28 'Bunker Buster') was developed for use against deep hardened Iraqi bunkers in the first Gulf War, bunker busting was the standard way of eradicating dug-in enemy soldiers, and it was no more sophisticated than using thirty-tonne vehicles to roll over and flatten enemy positions. Tanks were and are brute force killers, and on the Pacific islands, where Japanese tanks were both rare and useless (American M3 Grant tanks far outgunned Japanese light armour), bunker busting was a full time business. In Vietnam, American tanks came to be used as armoured bulldozers, with M48 Pattons directed to squash North Vietnamese Army emplacements (usually, but not always, abandoned during the day) and collapse deep earthworks. These industrial forms of muscle were the benign ones. More malignant demonstrations of metal on flesh came to light as the war proceeded. One soldier trapped in a tank momentarily stuck over an occupied Japanese foxhole determined, through his panic, what to do before the surprised Japanese attacked him first: 'And so I sat there in the driver's seat, not able to see, with two Japs under the tank ... I pulled the tank forward a few yards, then put it in reverse with one track locked. The tank spun halfway around crushing the two Japs and their machine gun' (Gilbert 193–4). The most direct way to use a tank is to flatten a human body with it (such a method is *so* vivid that it gave the unknown Chinese man who resisted a line of armour in Tiananmen Square in 1989 his short-lived power over the machine). Using a tank to squash a human makes it seem too outrageous to be possible, let alone commonplace.

Just as Germans became frustrated enough to use Russian bodies for tank decking, Americans, too, became inured to attacking human bodies with thirty-tonne vehicles. One Marine tanker in the Pacific recalls rolling over Japanese bunkers: 'You can hear them popping. Heads cracking, screaming, everything else ... You're laughing from ear to ear while you're doing it. I don't know what got into us at that point' (Gilbert 61). Explanations fail before the soldier's combined terror and relief, a frenzy at the enemy's weakness. The tank's ponderousness and speed, the concomitant terrors of being protected yet imprisoned by enemy fire, of being safe and also endangered, produces a hysterical rage against those outside. Human jaws open and the power of the tank slides in: then the tank closes up and grinds down the enemy. A veteran from the first Gulf War recalls seeing his wingman (the tank guarding his vulnerable flank) 'crushing bunkers and antitank weapons beneath his tread. His turret was traversing smoothly right and left, discharging main gun rounds every ten seconds' (Vernon 221). The tank fights above and below. The Abrams' suspension system and stabilized gun mount allows it to fire and move over uneven terrain (ground formed by human bodies in bunkers – some estimates suggest that more than six hundred Iraqi soldiers were buried alive by American armour) simultaneously. There's a slick machine quality to the language, the machine operating exactly as designed, the main gun going off at the highest fire rate a human loader can manage. The flesh body becomes only another brief hesitation along the way: 'Anderson drove over [the Iraqi]. The tank seemed to hit a speed bump, the treads presumably turned the man to a bag of white pebbles, white gravel, the tank rolled on. "You got him," said Burns. "How do you feel about it?" "It was awesome," said Anderson' (Sack 144).[20] The narrator has already described the scene in order to highlight the almost inadmissible pleasure of crushing a human with a sixty-eight-tonne tank, but at the last moment he sanitizes the image, converting what would be a messy wet red mash to its antithesis, orderly, dry, white, discrete items. What is thrilling must be offset by the innocence of the backyard patio where decorative gravel comes in bags. When the human is reduced to a bump that the crew barely notices, the tank body has almost entirely replaced the self. In the driver's mind there is no connection between the Iraqi and him: it's a neat trick, and then it's over. The patio has been smoothed out. The physicality of the shattering egg skulls the veteran from the Pacific talks about, nervously, indicates that something dreadful has occurred. But by 1991, that hysteria seemed to have settled into policy: run over the bunker, bury the enemy alive, it's a (live) speed bump.

Armour begins to take on more of its own life and nature. Tankers understand that, while they crew the machine, in some way they also *are* the machine, its weight and protection. Armour veterans from the

Second World War and after recount the process of friendly tanks purposely firing on each other with their machine guns: 'The tank's machine guns would protect each other. In fact it was great to have the Japanese swarm all over a tank, and the other tanks could just shoot them off with their machine guns. I remember hearing Giba yelling through his radio to his platoon leader ... [who] was very calm and cool, and he said to Mike, "Mike, don't be excited. We're killing 'em as fast as they show up"' (Gilbert 134). It requires an enormous amount of calm to bear having enemy soldiers on top of the tank, and then to have them knocked off by deliberate friendly fire. The tanker's panic at having enemies on his second body's skin brings out the platoon leader's domestic calmness as he reassures the other man. Shooting enemy soldiers off friendly tanks became so commonplace that green soldiers in Korea and Vietnam were schooled by veterans in the practice of 'scratching one's back' (Zumbro 12). Tankers, especially in the close fighting of the Pacific islands and Vietnam, regularly had cause to 'scratch each others' backs' in order to get rid of the pests lodged there. Vermin irritating the tank's thick skin would be scraped off with .50 calibre nails. Not only do the tankers learn to perform this task, they invite it. The enemy's bug body becomes a thing of ridiculous weakness; attacked by a persistent Vietnamese sniper dug well into a bunker, Ralph 'Zippo' Zumbro's tank captain orders,

'Holt drive us in there and neutral steer – we'll grind his ass out.' But to our surprise, the tracks only lifted slightly and didn't dig in.

'I got an idea, Sarge.'

'Okay Zippo, let's hear it.'

'We can pop him out like a zit. If we can get one track up on that bunker next to the hootch, I can get enough depression to put a delay shell *under* him.' A couple of minutes later we were in position; and – Wham! ... just in time to hear an infantryman yell, 'Hey, look at supergook!' The pressure wave of the detonation had launched the sniper like a human cannonball in a steep parabola, and his body disappeared over a distant roof. (Zumbro 40)

The fifty-two-tonne M48 Patton is an enemy crusher, acne cure, and a product of the Acme Corporation, supplier of Wile E. Coyote's gadgets. The amazing fact of laying a 90 mm high-explosive shell under one human being is indicative of the technological overkill of warfare, particularly armoured warfare, in Vietnam. When tanks are called on to perform heavy duties like killing other tanks, ramming emplacements, breaching walls, firing at targets kilometres away, they seem to make a certain kind of battlefield sense in Guderian's terms. But to aim the machine and its armament at one human with a rifle points to the insanity of the entire American mission in Vietnam, the colossal expenditures not only of *matériel* but of human beings, of the full force of the Acme Corporation that is all too happy to fill the coyote's crackpot orders (Zumbro's account has the flavour of a cartoon where Wile E. Coyote flies over a roof or cliff – it's good slapstick as long as you're not the target and the target isn't a real person). The lone soldier facing a tank cannot hide, and so he must either escape the battlefield or find some resources that make it 'survivable.' The human body re-enters the gun-armour spiral and begins to fight tanks single-handed.

Because they took armour seriously, the Germans designed the first, and in many ways most successful, hand-held anti-tank weapons light enough for the infantry to carry. The *Panzerfaust* ('armour-fist') looked something like a toilet plunger, except that the plunger end was full of TNT shaped so that when it hit the tank, a jet of superheated gas would shoot through the armour. The *Panzerfaust* was mounted on a hollow tube that guided the back blast behind the user (usually a *Panzergrenadiere* prepared to get within a suicidal thirty metres of a tank in order to fire at it). When the Allies saw German tanks in action, they too developed anti-armour infantry weapons. The British invented their much-loathed back-breaking PIAT (Projector, Infantry, Anti-Tank), crippling because of its badly designed cocking mechanism (the PIAT's round was launched by a large spring, instead of a rocket), and the Americans produced the Bazooka. These weapons shared the same basic qualities: they were portable, could be crewed by two, one if necessary, were recoilless (relative to rifles or machine-guns where the soldier's body absorbs the recoil), fired shaped charges that could blow holes in armour, and could attack tanks in their weak spots (the top, side, back): a determined, properly armed, tank-killing infantry soldier is someone to be feared. Second World War anti-tank weapons were rough, hard to use, dangerous to the infantry (some German tubes had '*Achtung! Feuerstrahl*' [Attention! Fire jet] printed on them in order to warn forgetful or untrained users that a smoking gas tail was about to spurt out the back of the launch tube), but they looked ahead to the current world of 'stand-off' weapons that can be fired from two or more kilometres away.

There were other even less enviable ways for exposed bodies to attack tanks. Japanese soldiers famously sat 'in holes or craters in the road, a 100lb aerial bomb between their knees and a brick in one hand ready to strike the exposed detonator the minute a tank passed over' (Perrett 159). The tank's destruction, true to a lesser degree with the *Panzergrenadieren*, seems to require the soldier's forfeiture of the body: on the Pacific islands, Americans uncomprehendingly watched Japanese throw mines on tanks, and themselves on top of the mines, in order to stop armour. All infantry learned that they needed to come perilously close to tanks in order to do them damage, that the flesh body is effective against the metal self only when it can make contact with the armour. Russian soldiers on the eastern front, never properly equipped with anti-tank weapons, learned to bundle grenades together and force them under tank turrets (assuming they could come close enough). Grenade packs, called satchel charges, would ideally blow the turret upward, lift it off its ring mount, and force an explosive blast into the crew compartment. Soldiers soon stopped talking about killing tank crews and instead focused on killing the tanks themselves. For the Germans, a T-34 with its ring-mount destroyed was the real goal of a flesh-on-metal attack. Such close-order tank killing was lessened by the next generation of shoulder-fired anti-tank guns. The Soviet RPG-7 (rocket propelled grenade launcher) supplied to the North Vietnamese made convoy duty in Vietnam that much more unsavory for mechanized troops. The RPG-7 could punch into 30 cm of rolled homogeneous armour and could be fired from 300–500 metres away. The RPG-7 also forced armour to turn against itself.

American soldiers in Vietnam didn't sandbag and ride on top of their tanks or APCs because it was hot (the Pacific islands were no cooler during Second World War). Buttoned-up armour hit by a rocket or mine would necessarily contain the full force of the explosion; not only would the rocket pierce the armour, cause the inner layer to spall, and light up unfired ammunition, but the warhead's explosion would consume the air, suffocating the crew. Then the fuel would catch fire and melt the vehicle. But, if the crew were riding outside and the hatches were open, while a mine or RPG might blow off a track or even set the vehicle on fire, the same explosion would also likely throw the crew off the armour. Experienced armour soldiers learned that buttoning up was a death sentence (Birdwell and Nolan 182). When the Soviet RPG-7 entered the gun-armour spiral, the gun gained ascendancy. Armour was still useful but became more limited: armour crews couldn't step away from tanks entirely, but the machines had to be modified. The easiest way to accomplish such alterations was to leave the vehicles open and ride on their tops, then to sandbag the tops in order to make tanks on tanks. The answer to the problem of increasingly powerful stand-off weapons has been to strengthen armour, but even Chobham Armour, the mystery agglomeration that protects contemporary tanks, is hard put to deflect recent anti-armour weapons.

Contemporary anti-tank missiles run on information technology and irony. There are still, from the days of the *Panzerfausts* and Bazookas, recognizable launch tubes and missiles: the rest has changed. At the end of its war in Vietnam, the United States fielded the TOW missile (Tube-launched Optically-tracked Wire-guided missile). The TOW could be used accurately up to distances of 3,000 metres (outdoing the Russiàn RPG-7 by a factor of six or more); thin wires attached to its fins allowed the soldier to guide the missile directly to its target (following the missile by eye meant that it was 'optically tracked' but the target had to remain in the line of sight [LOS]) – it was a true stand-off weapon. The TOW design was immediately copied, and current anti-armour missiles resemble their progenitor except for their infinitely heavier, more powerful, more lethal qualities. The ERYX and MILAN missiles (figs 8 and 9) are both made by a multinational corporation known as MBDA (Matra BAe Dynamics Alenia), the world's second-largest missile builder next to Raytheon.[21] The MILAN and ERYX have rendered survival inside a targeted M1A2 Abrams, Merkava Mark 4, or Centurion 105 an unlikely prospect.

8   **Mobility, of a Kind**: The ERYX missile with one user: mobility is an increasingly rare commodity.

9   **Loaded Down**: The MILAN Mark 3 with two users. Put tracks under it and some armour around it, and the soldiers are back in the box.

The MILAN is the most powerful of the two munitions but, because of its back blast, cannot by fired in an enclosed space (a room in a building) without baking the user. The ERYX has been designed for the tighter places in which architects of the Revolution in Military Affairs believe future battles will occur (they will be MOUT: Military Operations in Urban Terrain). The missiles have a range of between six hundred metres and two kilometres and can penetrate up to one metre of rolled homogeneous armour; both can defeat any tank currently in the field. They are undeterred by cheap battlefield tricks like fires or the electronic countermeasures a tank might launch: 'The firing post computer takes a "snapshot" of the scene at the moment the gunner presses his trigger, and this snapshot is constantly compared with what the sight is currently seeing. Should a pyrotechnic flare be fired by an enemy tank in order to confuse the gunner or the missile, the computer sees the difference between the present image of the flare in the computer circuitry so that as far as the correction computer is concerned the flare does not exist' (Hogg 182–3). The missile obtains a grip on its reality, the 'snapshot' of the target, and keeps that and only that in mind as it flies on. There will be no way to jam such a missile unless the onboard computer can be hacked while the missile is on its four-second flight. The missile not only sees and compares: it knows. The irony of the MILAN and ERYX is that they no longer look as if they can be controlled by two soldiers, let alone one. They are weapons bulked up on industrial war steroids, massive cowls that swallow their users, the way early camera hoods absorbed picture makers. The missiles require two men to carry them and a solid place for a launch site. They might be the sort of armament future infantry in powered suits will carry, but at the moment they put the soldier inside another, narrower, triangle of weight, power, and mobility. And, while they can punch through a metre of armour, they also dwarf soldiers, yoking them to their personal warheads.

Tank-killing missiles are portable in the way that 1960s television sets were: they can be carried only if there's no wheeled dolly available. The missile ensembles are field-warrior office cubicles and must be equipped to bear the computing power, optics, range-finders, and extra rounds, to say nothing of the power supply, required for their operation. Little wonder that manufacturers' missile-demonstration videos use an abundance of purposely shaky video clips and low angles showing soldiers lifting the missiles and running fairly long distances with them: the burden of proof lies on the weapons makers' *cinéma verité* (it's true that the overall systems, while bulky are relatively light: twenty-six kilograms for the ERYX, nineteen kilograms for the MILAN). Long-range stand-off weapons rely on soldiers firing from relatively invisible positions where there is sufficient time to arrange the launcher and program the missile. The weapons are no longer hand-held, nor are they artillery pieces, nor are they enclosed light armour: they are weapons from newly appearing liminal zones on the battlefield where vehicles make too obvious targets but unarmoured humans cannot survive. Since the American military is preparing for MOUT, among other kinds of Operations Other Than War, the demand for a missile with the same clout but less back blast brought about the ERYX, which MBDA promises won't bake the firer along with the target (the warhead launches more slowly than the MILAN so there is less of a fireball shooting back at the soldier).

The ERYX uses a 136 mm shaped warhead (remember that the formidable M1A2 Abrams has only a 120 mm cannon by comparison) and can be loaded and fired at the rate of five missiles every two minutes. The MILAN has a 115 mm shaped charge that can be fired at a rate of three to four rounds per minute. The missile grail is a 'fire-and-forget' weapon (also known as 'launch and leave' in the military's fond alliterative rhetoric) that will surpass the TOW missile's LOS or the MILAN and ERYX's Semi-Automatic Command to Line of Sight (SACLOS) guidance systems. The soldier will 'paint' (mark) the target, fire the missile, and run: the ammunition does the rest. In a final ironic turn, it seems that the best soldier-portable anti-tank weapon will actually be fired from a 155 mm howitzer artillery piece. For more than ten years, the U.S. Army has invested in 'Brilliant' Anti-Tank Missiles, or BATs, now joined by the Sense And Destroy ARMor missile (SADARM). In each M898 SADARM canister are two warheads that, when their sensors alert them, pop open parachutes and begin a slow, circling descent. As they circle, they scan for what they have been taught to recognize (by noise and heat signature) as targets (tanks are their first choice). Having selected a target, the munitions' rocket motors fire and they go straight at the tank's weakest surface: its top (the SADARM is one of a family of munitions sometimes called 'self-forging fragments' – the rocket motors' high velocity fuses the warhead into a dense penetrator). If top attack becomes as prevalent as is feared, armour will have to thicken there too (for that reason, future tanks may have no turrets). Putting SADARM aside, the move to relatively portable high-explosive stand-off weapons like MILAN and ERYX has reinserted the soldier in the gun-armour spiral. As anti-tank missiles gain in power, range, and smartness, tanks will once more be forced to reshape themselves.

Armour has always meant weight. Because the Abrams is so heavy (seventy to seventy-two tonnes at its worst), it has fallen out of favour with those continuing the Revolution in Military Affairs. The Abrams, Bradley (troop carrier), and M109A Paladin (self-propelled howitzer) are the celebrities of a vanished Warsaw Pact environment; only the undeclared desert wars (Desert Storm [the first Gulf War, 1991], Enduring Freedom [Afghanistan, 2002], Iraqi Freedom [the second Gulf War, 2003]) have allowed heavy-killing machinery the space it needed to perform (American soldiers referred to the apparent wastes of the Middle Eastern desert as 'wall-to-wall fuck all'). If Military Operations in Urban Terrain or Operations Other Than War are the future, as the RMA predicts, then gargantuan land battles will be replaced by asymmetric contests, David-and-Goliath fist fights between terrorists (or freedom fighters, depending on your politics) and superstates like the United States. In 2002 the attack on fat armour peaked when, after years of development, the Crusader project was killed by Secretary of Defense Donald Rumsfeld. Crusader was a two-piece war set, a self-propelled howitzer with matching resupply vehicle (fig. 10). Self-propelled howitzers are not new: they drive to their desired location, anchor if necessary, and begin firing. The M270 Multiple Launch Rocket System (MLRS) that proved to be so telegenic during night firing in the Gulf wars (the MLRS fires with a distinctive 'ripple,' not unlike the Second World War Katyusha rocket, or 'Stalin Organ' – a bank of rockets fired from the back of a large truck bed) is similar to a self-propelled howitzer. If targeted by the enemy, the Crusader (a tracked vehicle with a crew of three) would move on: no need to become mired in artillery duels – the enemy won't be able to locate the Crusader.

The original Crusader was one of the broadest, heaviest, land weapons on the planet. The crew selected the target, rate of fire, and timing, but the machine calculated the rest, putting the right amount of propellant behind the shells, all of which are handled by an auto-loader. The Crusader fired ten 40 kg rounds a minute for three minutes (slightly slower when it was on the move). One of the Crusader's promotional videos showed it firing fifteen rounds in eighty seconds, while the experienced crew of an older, human-loaded, M109A Paladin self-propelled howitzer takes nine sweaty minutes to accomplish the same job. Because the amount of propellant can be altered with each charge, the Crusader was capable of dropping eight shells on the same target simultaneously. The Team Crusader website, once hosted by the parent company, United Defense, dubbed these special artillery moments MRSI (pronounced 'mercy') missions (Multiple Rounds, Simultaneous Impact). My grandfathers, one who fought in the trenches and one who was on the way when his troop ship was put into quarantine, would have known what to call a MRSI mission: a 'stonk.' It is the unexpected arrival of complete artillery devastation: any live stonk victims would be so grievously injured that they would presumably require mercy-killing. The Crusader could throw a 40 kg shell (including the SADARM canister) forty kilometres, earning it the name of a 'reachback' weapon because it reaches deep into enemy lines. The crew, protected by the usual NBC over-pressure, would sit in an armoured cabin that resembled more a jet cockpit than a tank turret. It had somewhat less mobility and speed than an Abrams, but it was pretty nimble for a sixty-tonne vehicle (39 kph across country, 67 kph on roads). The Crusader also came with its own prefabricated mythos including a fake heraldic shield complete with crossed swords, apparently dreamed up by the same United Defense savants who coined the 'MRSI' phrase. The crest's emblems were further explained this way: the lion represented 'Richard the Lion-hearted, King of Battle' (another ironic choice for Americans), a lightning bolt for 'digitized communication,' a horse head and wheel as signs of mobility and

10   **Twice as Heavy**: Crusader with its automated resupply vehicle. Total weight? More than just about any aircraft in the world can carry. Without the resupply vehicle and firing at maximum, the Crusader would be a lame duck in a matter of minutes.

resupply, and a stack of ten cannon balls for the Crusader's battle-field prowess.

For RMA adherents, the Crusader represented the military's worst gourmandizing habits. It and its companion fully automated resupply vehicle required, if not transported by ship, the largest aircraft in the world (the Lockheed C5 Galaxy or C141 Starlifter) to reach the bat-tlefield. Even after then-Army Chief of Staff General Eric Shinseki pared the Crusader down to forty tonnes, it was still hard to guarantee that the planes would be sufficiently available for the howitzer dur-ing wartime. The program's cost struck the RMA's backers as equally Rabelaisian: more than $11 billion USD from start-up to delivery. Before the election of 2000, George W. Bush attacked Crusader as the kind of military boondoggle he would raise off the taxpayers' backs. For a while, in part because Crusader was interoperable, because it matched the heavy-army weapons suite (Abrams, Bradley), and because United Defense had the right kind of contacts on Capitol Hill, the program was sustained. But once Donald Rumsfeld's hold on military operations was secured in the post-Afghanistan White House, General Shinseki found himself publicly kneecapped. Shinseki, who had overseen Crusader from inception in 1994, who had risked his career to drag the army toward the Joint Strike C4ISR world, who had pushed the unwilling Army to move from heavy to light forces ('If you don't like change, you're going to like irrelevance a lot less,' he told his subordinates [Boyer 54]), now found himself shoved in front of a political tank when Rumsfeld, without telling Shinseki, stopped the Crusade.[22]

While heavy armour is as attractive to the current ground force as horses were to the cavalry a hundred years ago, the Army has begun to field its interim light force, the Stryker Combat Brigade Team (SCBT).[23] All SCBT vehicles in their ten different configurations (wheeled, not tracked) use the General Dynamics Land Systems' (GDLS was GM Defense until April 2003) Stryker chassis. Despite the funding of new armour, there is a sense of grief in the so-called tread-head community over the loss of the Crusader, and fear that, with the Abrams acquisition cycle complete and no visible plans for the manufacture of future supertanks, *real* armour is finished. David Drake, who served with a mechanized unit in Vietnam and then returned to the United States to write a series of highly popular science fiction stories about a heavy armour mercenary corps, suggests that portable fusion power would release tanks from their burden, allowing them to float over the battlefield on air cushions produced by ducted fans. In 1974 that was science fiction. In 2001 Patrick Wright was told at the Army's Fort Knox cybertank centre (a compound with over a dozen networked tank simulators) that the future tank might not 'even touch the ground – maybe it will ride on an electro-magnetic cushion; maybe it will have no turret; maybe its gun will be a laser beam' (431). Dreams of new power plants and suspension systems indicate that the Army won't easily surrender its main battle tank.

The SCBT is proleptic of the Future Combat System (FCS) about which RMA leaders dream. Like all components of the RMA, the FCS is a system of systems. Service priority (Army, Navy, Marines,

Air Force) is assigned, so RMA interoperable theory goes, by the military goal to be achieved rather than by the pride of the forces involved: Joint Force operations are intended to set aside decades of interservice infighting. The FCS is 'an overarching effort to field by 2010 a family of up to 18 types of lightweight combat vehicles and robotic platforms, all linked under a common command-and-control network. The project is the cornerstone of the so-called transformation of the Army into a lighter and more mobile force' (Erwin 'Army "Transformation"' 22). There are no real surprises here since lightness and speed are currently the popular sides of the armour triangle. But United Defense, corporate parent of the dead Crusader, is at work on other currently classified elements of the FCS; immediately after the Crusader was flattened, United Defense 'joined' with General Dynamics Land Systems in a lucrative ($4 billion USD over six years) contract for the next generation of armoured vehicles.

A large part of the Future Combat System is occupied by robots. Even before Kennedy and Johnson's secretary of defense, Robert McNamara, explained his 'McNamara Line,' an electronic surveillance fence that would partition North and South Vietnam, the military invested in war robots. The 2002 defense budget allocated $500 million USD between 2004 and 2009 for ground robotic vehicles called UGVs (Tiron 'Lack' 33). DARPA has already planned similar systems, now entering a second design phase, such as self-healing minefields, where networks of mines talk to each other and monitor their own health. If a mine, or belt of mines, is destroyed, the survivors shuffle on the battlefield (represented by DARPA as an electronic chessboard), hopping (not rolling on wheels or treads) into new positions so that the breach is closed. The newly allocated $500 million will go to solving machine-perception problems that stand between robots slaved to humans and fully autonomous robots. The RMA, which the second Gulf War helped brand in the public mind as an example of the new army's success, puts UGVs out in front, and Congress has followed, 'mandat[ing] that at least one of every three future Army systems be unmanned' (National Research Council *Technology* 1).

UGVs are graded by the degree of human participation they require to function, from the full involvement of humans using joysticks or graphical interfaces to operate teleguided machines to completely autonomous hunter-killer robots. There are four groups of UGVs: searchers (sent into danger to observe, lift objects, explode un-exploded weapons – iROBOT's 'Packbot' is similar to the searcher); donkeys (wheeled carriers carrying infantry supplies and packs, fol-lowing trails of electronic markers called 'breadcrumbs'); wingmen (nearly autonomous robots that operate as infantry sidekicks); and hunter-killers (armed robots that can fight, replenish and repair them-selves, then stand down for a period of thirty days, reactivating them-selves when necessary). Each of these machines is proportionately distant from us in time and engineering capability (the predictors say autonomous vehicles will arrive between 2015 and 2020). Imagined hunter-killer teams would 'consist of at least 10 medium-sized "killer" unmanned network-centric autonomous ground vehicles. Each killer vehicle would carry internally (in a "marsupial" manner) up to five small network-centric autonomous "hunter/observer" ground vehi-cles' (National Research Council *Technology* 27–8). Nothing on the battlefield will go unconnected: the systems are interactive and com-municative; killing machines will have pockets full of searchers that will gather intelligence and relay it back to them. Marsupial robots (as contemporary videos show) work in an eerily organic fashion, unload-ing from each other slowly and deliberately as they lower themselves onto the field.

Teams of autonomous vehicles ranging across the battlefield to attack a presumably under-armoured and under-gunned enemy will be part of the future, if RMA policies hold. The implications for humans are strange enough that I have set aside a chapter in order to deal with uninhabited vehicles on their own. No UGV supporter would suggest eradicating heavy equipment like Abrams tanks that are needed to do specific crushing work on the ground; if UGV con-tractors want to stay in business, they must continue to talk about uninhabited vehicles as being helpers in a networked team, slaves (a translation of the Czech word 'robot' after all), to human soldiers in future combat systems. Dedication to armour for its own sake is a trap, says this RMA spokesman: 'What is at issue is therefore not helping a particular weapon, however tradition-laded it might be, to score a measure of success, but of winning the battles of the future, and winning them so completely, so speedily, and so comprehensively that they will bring the war to a rapid conclusion. All the weapons must co-operate to this end, gauging their own performance and demands according to those of the arms that have the greatest striking power' (205). The author subscribes to all the components of the RMA: speed, shock, communication, interoperability. The human body is shoved aside by the weapons themselves, full of grim agency, capable of assessing themselves. It is the weapons that will finish war, weapons that will work together, weapons that will ignore limited victories and achieve a definitive end. Agency slips from human hands and into machine bodies. The above author is not Admiral Bill Owen (*Lifting the Fog of War* [2001]) or Harlan K. Ullman and James P. Wade (*Shock and Awe* [1996]) – but rather Heinz Guderian – Heinz Guderian looking ahead to a war that he has not yet fought.[24]

In Vietnam, exhausted from eleven hard months as a decorated tank commander, Dwight Birdwell finally succumbed to his friends' pleading that he take a relatively safe post in a firebase until his tour was up. Deep into post-traumatic stress, drinking hard and sleeping light, Birdwell lived in his hootch, waiting to fly home. One morning he awoke with a terrible feeling that 'I needed to get out of [my hootch], something was going to happen – and, bam, just moments after I dragged myself outside, a big old M48 tank noisily crashed through the hootch, crushing my cot.' But the main thing is: 'There was no driver in the tank' (Birdwell and Nolan 188).

Chapter Four

Rotor Hearts:
The Helicopter as War's Pacemaker

Airmobility, dig it, you weren't going anywhere. It made you feel safe, it made you feel Omni, but it was only a stunt, technology. Mobility was just mobility, it saved lives or took them all the time.

– Michael Herr, *Dispatches* (1977)

My job was to fly over Vietnam. I didn't walk in it and didn't want to go down there.

– Craig Matlock in *Headhunters* (1987)

When we fly, soldiers don't die.

– 1st Air Cavalry Colonel Jim McConville (2004, Iraq)

Blades thumping rhythmically, the Bell UH-E1 'Huey' Iroquois helicopter, sign for America's presence in 1960s Vietnam, flies across the abyss between the late mechanization of the Second World War and the current information war in transparent battlespace. The gas-turbine-powered Huey turboshaft rotor marks an early stage of what will become the Revolution in Military Affairs, as America once again found itself through, and in, high technology. The rotor that turned lazily during the early days of battle in Vietnam is running at full throttle now, holding aloft the Boeing Sikorsky RAH-66 Comanche stealth helicopter as it flies backward, banks vertically, slips sideways, all at high velocity, sliding fluidly through C4ISR battlespace. The helicopter operates at the edges of things. It hangs between the battlefields of the Second World War and Vietnam; between the technology that ended the Second World War and that which now drives fully interoperable so-called 'netcentric' warfare organized around information dominance (as well as full connection to the GIG); between the ground and the lower deck where the Air Force usually flies – like most hybrids, it is a little of everything without being any one of them. The helicopter arrived before Air-Land doctrine, the precursor to contemporary military understanding of the way air, land, and sea forces should interlock during combat; the machine had an early role as guerilla fighter – it took its enemies, and even its users, by surprise. The helicopter gained sudden entry to impassable war zones, brought wounded soldiers who would have died in past wars to sophisticated forward surgical hospitals, and resupplied an army permanently in the field. The helicopter is another moment of technological amplification and extension of the human body, which, like other such extensions, causes panic. For all their power, the machines are surprisingly vulnerable, requiring hours of ground care for each hour of flight. They bring with them a world of support: whole armies of technicians and specialists must be on hand to service the machines.

For a long time helicopters, first used by the Germans at the end of the Second World War and then by the Americans in Korea and the French in Algeria, were limited in range and lift by their reciprocating engines. These small motors were unreliable and unsuited to flight, producing vibrations that made the rotorcraft hard to fly, almost impossible to control. When Bell Helicopter's turbine-powered UH series appeared in the early 1960s (the name 'Huey' came from Bell's terminology: Utility Helicopter, series E, number 1, or UH-E1), it quickly came to dominate the helicopter industry. Bell outfitted the Hueys for action in Vietnam in a number of ways: as troop carriers ('Slicks'); attack ships ('Guns' as well as more heavily loaded gunships called 'Hogs'); medical rescue ('Medevac' or 'Dustoff'). Beside the Huey flew the small, quick, two-seater reconnaissance Hughes Aircraft OH-6 Cayuse Light Observation Helicopter ('LOH' or 'Loach') and, late in the Vietnam War, the Bell AH-1 Cobra, a tandem gunship. Hueys were used at ground level but also at 900 metres as Command and Control decks for captains and majors directing action on the battlefield. Reporter Michael Herr, who found himself obsessed by the men, the war, and Vietnam itself, fell in love with the helicopter. He saw in them 'nimble, fluent, canny' war participants that appeared to be 'death itself' (Herr 9). But the helicopter's strengths as a weapon of speed and flexibility could be erased by commanders who used them in routine, orderly fashions (Doleman 38). The wrestling match between the inherent chaos of helicopter warfare and the state's wish to routinize the battlefield continues as new generations of helicopters sacrifice mobility for heavy armament. As with other military-technological concerns (personal armour, tanks, UAVs), it is not clear which will be supreme: the human or machine, the organic or the rotor, heart.

The helicopter of the 1960s was an unlikely war tool. It was delicately structured with an almost papery aluminum skin, its power plant and hydraulics required endless hours of tuning and refining, and its rotorcraft body was covered with exposed fuel lines and connector rods, as if the ship's organs had been everted, arteries and tendons all visible, vulnerable. But the helicopter's fleetness, its ductility, match it well with battlefield chaos, what Clausewitz famously called the friction of war (119). Clausewitz likened accomplishing objectives in war to 'movement in a resistant element' (his analogy is to walking in water): the helicopter was thought by its early theorists to be a force that could eliminate that resistance (120).[25] Although it is neither jet nor tank, it can act as both, deliver high explosives and rockets, move over impassable terrain. Robert Mason became an experienced helicopter pilot at the beginning of the U.S. military's heavy reliance on rotorcraft in Vietnam. He began his training, as do all pilots, on the ground. But once in the Hiller trainer (because of the shortage of Hueys, pilots were trained in H-23 Hillers, helicopters powered by reciprocating engines rather than turbines), he found only a frenzy of events: 'My senses were overwhelmed by the clamor and bouncing and vibrations of the H-23 [Hiller piston-engine helicopter]. The blades whirled crazily overhead; parts studied in ground school in static drawings now spun relentlessly and vibrated, powered by the roaring, growling engine behind my back. All the parts wanted to go their own way, but somehow the instructor was controlling them, averaging their various motions into a position three feet above the grass. We floated above the ground, gently rising and falling on an invisible sea' (Mason 31). The machine appears to be chaos reified, an object at war with itself. The helicopter demands the user's complete attention: both feet control pedals that direct the rudder and stabilizer; one hand holds the cyclic, a sort of combined joystick, rotor-control, throttle, and radio key, while the other attends to the collective control and cockpit switches. The cyclic itself is an early indicator of a computer joystick – complex, flexible, hard to master: once mastered, intoxicating, addictive.

The cyclic is a machine arm offering the pilot an open hand. Without grasping that hand, without knowing the feel of the cyclic – there is no time to look down at it – the pilot cannot fly the craft. Through the hand, the arm, and the rotator cuff that reaches through the machine's deck to its insides, the pilot and rotorcraft are intimately connected: the two-way grip makes the human body a flying extension. What makes all the forces balance is the pilot's calm mathematical direction, indicated by Mason's word 'averaging,' although even mathematical averaging isn't a sufficiently complex term. Flying the helicopter requires endless feedback between the two organisms: the human body recruits all its feeling to govern the craft, correct for breezes, hot and cold air, dust, vapour, as they push it around. Mason's sense that the craft is cushioned on a sea of air lets him experience air through the rotor's extension – here is Clausewitz's world of friction made palpable, if not visible. The powered helicopter is about motion and noise; without power, the helicopter, unlike the fixed-wing aircraft, becomes a barely controllable dropping rock.

American rotorcraft manufacturers like Bell, Sikorsky, and McDonnell (later McDonnell Douglas) pushed helicopter technology to meet the defence leaders', particularly Robert McNamara and William Westmoreland's, demands for battlefield air mobility (Coleman 6, Westmoreland 109). Bell's Huey provides a very different flying experience for Mason, accustomed to the Hiller-23 trainers: 'When the IP [Instructor Pilot] squeezed the starter trigger on the collective, the response was a shrill whine as the highspeed starter motor began slowly to move the blades, not the clacking cough and roar I was used to. At operating speed there was no roaring, vibrating, or shaking, just a smooth whine from the turbine. The IP signaled me to pull up the collective. The big rotors thudded a little as they increased their pitch, and the big machine left the ground like it was falling up' (Mason 44). The helicopter rises from the Second World War and Korea, freed by a turbine that spins out a solution both to rotorcraft design flaws and battlefield paralysis. Everything in Mason's narrative is about smoothness: the noise is a high-performance scream, the actions have a slick texture. U.S. Army reliance on the early Hillers vanishes in the vapour of high-octane technology (only a few years later, gunship pilot Dennis Marvicsin will reject the Huey in favour of the Cobra, comparing the two machines as 'a Volkswagen Beetle and a Porsche'[170]). Vietnam veteran Philip Beidler recognizes in the Huey an enormous narrative of 'performance in the war that got us there ... eventually hypnotized us there' ('Last' 54), a narrative familiar to the United States, which is 'by its nature a technological nation. The American regime is a technical contrivance intended to achieve an unnatural end – peace and tranquility. In the same way, technical solutions to the problems of war are as natural to America as bravery was to Spartans' ('Last' 17). Soldiers caught in industrial warfare have seen the United States send materiel instead of men. Second World War poet veteran Louis Simpson wrote nimbly: 'For every shell Krupp fired/ General Motors sent back four,' identifying that the real battle was between factories – the German Krupp steel and gun works that ruled weapons manufacture in Europe for over a 150 years, and the similar dynamo at the heart of the industrial United States, General Motors – not workers (45–6). The helicopter is the next ingenious device the American forces will use to stay over Vietnam but, ideally, out of it. The helicopter seems to allow for a kind of arm's length war.

Initially, air mobility is understood as the cavalry reborn (Mason talks about the helicopter as a horse-team that needs reining in [34–5]). The creation of the air cavalry (simply 'the Cav') is a marriage of technology and frontier mythos:

> The First [First division of the Ninth Cavalry] was to be the first cavalry division in history with total air mobility. Platoons of slicks would

> ferry the grunts into battle, escorted by platoons of gunships. Every-
> body thought they would be The Answer, because obviously, getting
> around in the jungle using tanks and trucks just wasn't cutting it. No,
> the First Air Cav, with its hordes of helicopters, and its ability to insert
> troops anywhere at any time, would certainly Solve the Problem. But
> meanwhile, things were beginning to get a little ghastly. (Marvicsin and
> Greenfield 54)

Rather than dig into, float down onto, or grind over the earth as the
infantry, airborne, or mechanized units must, the Cav executes 'insertions' and 'extractions,' indexes of McNamara's managerial war. Air
mobility seems to abolish the earth. Hovering tanks armed with fourteen rockets, two mini-guns firing 6,000 rounds per minute, and two
M60 side-firing machine-guns guard air trucks disgorging infantry.
Gunship firepower ratchets up when, in 1968, the Huey Cobra and its
two 1,500-round mini-guns, 5,000-round electrohydraulic mini-gun,
40 mm cannon and 76 rockets, floats into Vietnam (Marvicsin and
Greenfield 199, Spalding 55–6). Unlike fixed-wing pilots, helicopter
pilots rarely fly higher than 450 metres, choosing instead to fly nap-
of-the-earth at 30 metres or less. The pilots are 'hot-rodders ... Flying
along the rice paddies with the skids almost in the water, raising up
over paddy dikes and hedgerows' (Ingram in Brennan 137). The
image of horse cavalry realized on the Cav's famous shoulder patch
(fig. 11) becomes a veneer for the mid-century remythologization
of masculinity: the hot-rod – speed, virility, danger, prowess, and
romance, assembled at home by the mechanical whiz kids in the
garage – is the true signified.

Three apparently diverse strands need to be brought together. The
first is the daily reality of technological fixes on which wartime
industrialized nations are urgently dependent: as Philip Beidler says,
we can be hypnotized by the system that has worked before. Technology presents a paradox for RMA adherents: solutions to war problems have been and will be sought through harnessing of the laws of
physics (the atom bomb), chemistry (napalm), mechanical or aerospace engineering (the helicopter). Overlying the expectation that
technology can save us is second layer – the myth of the frontier. No
matter how much high technology the state or its people use, there

11   **Horse of a Different Colour**: Air Cav shoulder patch – still a horse, even if it's now a flying one.

persists the belief that the strength of will that made it possible to drive wagon trains across thousands of kilometres of hostile country operates unadulterated today. The open quality of the frontier proposed anxiety for Americans: How would they dominate that vastness? Certainly by horse first, but then more definitively by laying iron train tracks into the earth and with them writing a story of dominion. The Vietnamese jungle offered an apparently unconquerable depth: if one couldn't build a railroad there, and tracked train cars like tanks and APCs were too vulnerable, then perhaps having trackless air trains might work. Vietnam comes to be the new frontier that, like the wild west, will be tamed by technologically ingenious pilgrims with access to the latest tools. Finally, there is another connection between the two: the hot-rod builder, the kid with the home chemistry set, the amateur tinkerer and inventor who will become Thomas Edison, or Albert Einstein, draws strength from the myth of the single man, range rider or engineer, who can unite craft, mechanical ability, and the pilgrim's frontier heart.[26] It is no mistake that the troops called on to go forward into what the American army called 'Indian country' (land held by enemy Vietnamese) were largely drawn from Ranger battalions, lone-ranging soldiers who went ahead of the others. The horse, hot-rod, and helicopter share the some crucial elements, but the helicopter surpasses all. Its speed, ability to fly nap-of-the-earth, reversibility (depressing one pedal allows the pilot to turn the machine on the central rotor and fly in a new direction), power to erase predictability, even time, make the spinning machine a new player in the fluidity of battlespace, what pilot Robert Mason has already identified as that 'invisible sea' on which the rotorcraft floats.

As secretary of defense for presidents John F. Kennedy and Lyndon Johnson, Robert McNamara believed that Clausewitz's friction of war, and the accompanying chaos of battle, could be overcome by statistical modelling and managerial warfare (Gibson 80). The helicopter would play its part: the rotor would blow away the obscuring fog, while lifting troops off the ground would erase the friction, so that what Clausewitz called 'the great beam' could 'turn on its iron pivot' unencumbered (119). Seeing an ocean of air above them, an ocean in which a craft could tread 'water,' military thinkers began to craft new strategies for ground war. Until now, no machine could wait overhead, rising and falling like an elevator, yet land swiftly and accurately when called. There are no ramparts in the sky, what siege experts once referred to as the 'glacis,' shoulders of land used by fortress builders to break troop advances (the same term used, as we have seen, for the slanted front part of the rolling fortress that is a tank). Limitations from previous wars had suddenly been lifted. The ocean appeared to be without ramparts, freeing the strategist from the siege states that once organized war, particularly war understood by Marshal Sébastien Le Prestre de Vauban, a seventeenth-century master of siege warfare, fortification, and trenches (Vauban's theories were literally practised to death during the First World War). Helicopters fall up into a newly opened space. When it saw the turbine-driven helicopter in action, the war state grasped the notion of the open sky: 'Senator Henry Jackson was already enthusing that in Vietnam "The sky is a highway without roadblocks. The helicopter frees the government forces from dependence on the poor road system and canals which are the usual arteries of communication"' (quoted in Spark 88).

The helicopter tears ragged new holes in the air, rips different from those made by supersonic jets, long-range subsonic bombers (the B-

52 stratofortress), or lumbering aircraft used for close support fire (the AC-47 'Spooky' or, more colloquially, 'Puff the Magic Dragon,' now upgraded to the AC-130 Spectre).

The helicopter unravels the local sky, pulling on loose threads until it falls apart. The North Vietnamese Army digs – Americans fly: 'Under the ground was his, above it was ours. We had the air, we could get up in it but not disappear in to it, we could run but we couldn't hide' (Herr 14). The United States used hundreds, then thousands, of helicopters to unthread the sky, approach, avoid, leave dead Vietnamese behind and remove dead and living Americans. For all his love of the helicopter, Herr is no more taken in by its promise than Vietnam veteran Philip Beidler. But, for true believers, air-mobility means temporary occupation and departure, fluidity and solidity, the future and also the past, as military historian J.D. Cole-man indicates: 'The techniques developed by the 9th Cavalry Squad-ron – the scout ship/gunship combination ... the use of aero-rifle platoons to develop situations to the point where conventional infan-try units could be profitably committed – meant that the Army had a true cavalry again. This was a cavalry whose differential speed advantage over ground vehicles was analogous to the earlier differen-tial of a horse soldier over a foot soldier' (268). So anxious is Cole-man to understand new technology in terms of a past military culture that he blinds himself to fact that routines and formations that typify the horse cavalry commander's world-view eliminate the helicopter's advantages. The Air Cav organized helicopters to fly in groups called 'gaggles': formation flying meant that the otherwise highly mobile craft were now locked into predictable patterns. When ships landing a kilometre away reported taking fire, helicopter pilots were told to stay on station and put down in the landing zone as planned. If a patient enemy soldier on the ground didn't hit the first helicopter, he or she could rely on having a number of other chances as each craft systematically landed where others had just taken off. Instead of allowing the pilots to handle their own approaches, the Air Cavalry eliminated the confusion that could work to its benefit. Even for the circling of landing zones, Cav doctrine had pilots flying in predict-able cloverleaf patterns: Vietnamese infantry wishing to take down a helicopter didn't have to try to follow a particular machine – all they had to do was wait until another ship flew into the gun sight. For a brief time, the American forces held the advantage of an unknown, unpredictable, powerful, fast weapon, but the Vietnamese wouldn't always be panicked at the sound of helicopter blades – soon the soldiers would learn to shoot at the vulnerable parts of the craft (the spinning blade, the tail boom, its small rotor) and bring them down. Thinking about horses and helicopters together vitiates the new machinery's effectiveness: orderly formations remove the mobile from air mobility. In fact, to be successful with the Cav means acting autonomously.

The helicopter gives the illusion of freedom. It seems to provide for the quick drop off and recovery of troops, for darting attacks, disappearance into an open sky. But beyond a certain limit the military cannot tolerate disruption: it may need mavericks to fly for it, but it also cannot let those lone rangers do whatever they please. On the ground the Marines found their Vietnamese enemies to be effective, frustrating guerrilla fighters. There were almost no pitched battles, rarely were the Marines able to find Vietnamese dead after a firefight, and usually they had to settle for blood trails or drag marks indicating that one soldier had dragged another away. Vietnamese soldiers and revolutionaries tunnelled extensively, lived in deep bunkers and 'spider' holes, dug out underground warrens where they ran surgeries, slept, lived. But Americans could be as surprising and frustrating using helicopters, the technological equivalent of Vietnamese guerillas, dodging and disappearing, dragging off the wounded and leaving the enemy no satisfaction.

American military doctrine in the mid-1960s had not caught up to its own technology. It saw the helicopter as a new version of the horse, instead of a wasp or hummingbird. Military command kept making the same mistake, trying to use the helicopter in an organized fashion, quelling its flexibility, its penchant for swift, infinitely reversible, unpredictable combat. Where command might want order, the pilots and soldiers of the Air Cav knew better. One Cav soldier reports: 'Our tactics were speed and killing. We were always the aggressors. We took the initiative and would not hesitate to level a village at the slightest provocation. We used artillery and air strikes a lot, never took unnecessary risks, and interacted with the smooth precision of small units' (Borsos quoted in Brennan 158). 'Speed and killing' aren't tactics, but the strategy is clear: everything will be subordinated to that goal. Getting the job done is more important than the method – and repeating the method is unwise (these ideas are central to the joint interoperable warfare now being supported by a number of U.S. military professionals and theorists). The focus on small units speaks to guerilla fighting, and these Cav troopers were effective at it, as reporter Michael Herr attests: 'They killed a lot of Communists, but that was all they did, because the number of Communist dead meant nothing, changed nothing' (96). Behind the Vietnamese resistance was relentless will to self-governance, to be free of first French then American colonial power. Further, connected to the apparently flexible helicopter is an enormous technological umbilical providing fuel, parts, mechanics, munitions. Not only do the troops require unceasing fire support, but they must also return the machines to their bases for resuscitation. Ten thousand kilometres of supply lines mate the helicopter in Vietnam with the weapons industry in the United States. Underlying the trooper's narrative of apparent mobile power is an enormous static structure of fortified camps, maintenance crews, logistics, and the inevitable bureaucracy that keeps the whole moving.

The helicopter in Vietnam hangs shudderingly between the world of the early 1960s hot-rod enthusiast and twenty-first-century networked battlespace. From the world of the hot-rod comes an addiction to speed, to 'souping up' the flying car. The helicopter crew chiefs charged with maintaining the craft became Army shade-tree mechanics specializing in upgrading their equipment. One crew chief, Don Reacher, impressed Robert Mason as the master of his ship. Caught in heavy fire as the Huey lifts off, the pilot draws more power from the turbine than the Huey should be able to muster. Puzzled, Mason asks, 'How did you know this ship would be able to do that,' only to be told: 'Simple. This is Reacher's ship' (Mason 133). The pilot flies but the crew chief owns: Mason discovers that, like any mechanic thoroughly in love with his equipment, Reacher has 'made certain fine, illegal adjustments of the turbine' and produced a unique craft capable of more than the official specifications indicate (to his shame, Mason later crashes this ship, which has almost become a character in the narrative). The Cav pilots knew that routines on the battlefield signed their death warrants, and determined to avoid them: 'The problem with [flying helicopters in] cloverleafs was that the enemy could just keep firing their weapons and the ship would fly back into the bullets. Our swirling, screwing type of pattern prevented this from happening' (Paulmeno quoted in Brennan 286). Using unexpected manoeuvres, illegally modified turbines, breaking out of formation, all were ways crews found to retake control. Policy back at the base might be to have the pilots hold to predictable cloverleaf patterns (fear of mid-air collisions was very real, and establishing flight routes resolved that official worry), but the survivors tended to stay above small-arms fire and then charge for the ground in crazed 'swirling' dives. After a year of watching formation-flying kill his fellow pilots, Mason learned that the best way to act was erratically, surprisingly: 'When they called us, I was busting through the trees. I had swerved off to the side of the stand and then swung back in fast. This allowed me to bank very sharply so that the Huey and its big rotor disk squeezed between two tall trees thirty feet apart. After hurdling through the trees, I flared the ship quickly to make the landing. The radio operator who had been asking where we were said, "Oh." We landed right in front of the squad' (458). Mason is ahead of the infantry requirements. He's been in the situation before and knows that he must approach it as he never has – no action can be safely repeated. It's better to turn the craft on its ear than fly right into a hot landing zone.

Vietnam helicopter pilots discovered by trial in the 1960s what in the 1980s would become the military doctrine of AirLand Battle, a form of combat that, according to cultural theorist Chris Hables Gray, 'compresses ... real (i.e., lived, human) time. AirLand Battle goes twenty four hours a day, thanks to caffeine, amphetamines, computers, radar, and infrared. Rapid attacks, slashing and thrusting, shatter the enemy fronts and armies in a matter of hours instead of the days that blitzkriegs took, or the weeks and months of most modern-war offensives' (40). AirLand battle's celerity is founded on information technology coupled with *matériel* capable of arriving on time with the necessary firepower. A Huey pilot during his first tour of duty in Vietnam, Dennis Marvicsin was unsatisfied with the craft's lack of agility and firepower. He dreamed of a machine that would be faster, sleeker, better armed: shown the Bell AH-1 Cobra attack ship, he feels 'like he had been ushered into The Presence' (170). The Cobra sets a new level in unpredictability: 'You could dive it, you could put it into highly unusual attitudes, you could damn near barrel-roll it'; as for weapons, 'it had miniguns and a 40mm grenade launcher. ... It had twice as many rocket pods as the Huey, and they would fire high explosive, white phosphorus, fléchettes, chocolate pudding, whatever' (171). Marvicsin believes that this is the weapon they have needed all along to win the war, just as the hot-rodder focuses only on the track to the exclusion of all else. The Cobra transcends the category of machine to become a spiritual artifact that will bring the United States salvation in war. Amazement at machinery becomes adoration when that technology allows the human to do what would otherwise be absurd. Marvicsin wonders that 'in a steep dive, you could glance up through the fighter-style canopy and actually look at the ground through the disk of the rotor blades' (171). The Cobra allows the pilot to defy gravity, to defy even what he has come to expect of helicopters; unlike fixed-wing aircraft, rotorcraft are prevented by the physics of the rotor disk from inverted flying or barrel-rolls. Here is a religious experience of seeing the world, looking *up* at the ground, through the machine's agency. The pilot is trapped by speed, seduced by the compression of the Huey into the thin Cobra profile. As used by American forces in Vietnam, the helicopter produced the fastest combined air and ground war in history – not even the German Second World War *blitzkrieg*, usually the marker of extreme battle tempo, is comparable. Even before AirLand Battle, the helicopter cranked up the speed and delivery of weapons and troops: historian Alasdair Spark calculates that the Army 'mounted 2.9 million sorties over 1.2 million hours – or, put another way, compressed 137 years of activity into one' (90). Yet, for all this power, as military theorist James William Gibson points out, the 'highly routinized pattern of most search-and-destroy missions meant that the Vietnamese knew where American forces were' (104). The helicopter revealed the rest of the machine, would land in predictable places, and be vulnerable.

As with the human enclosed in powered armour or tanks, helicopter crews found themselves both exhilarated and terrified by their

12   What's the Buzz? The redesigned Bell AH-1Z SuperCobra, now with four blades but still the narrowest (1 metre inside the cockpit) rotorcraft, also one of the fastest. It can lift more, fly faster and farther than ever. It comes armed with a three-barrel 20 mm electric gatling gun that puts out 750 rounds per minute, and various rocket pods, Hellfire, or Sidewinder missiles as needed. Designed only as an attack helicopter.

machines. The pilot who survives has inevitably learned present-mindedness and, even when caught between enemy fire and the failure of delicate rotorcraft, will persist in struggling for agency. Marvicsin recalls sitting in his wounded Cobra, unable to get enough lift to escape attack, but where he doesn't have lift, he has spin: 'Maverick kicks the right pedal viciously and the crippled Cobra spins madly on its rotor axis, whirling like it's suspended from a skyhook, spitting minigun fire and rockets to all points of the compass' (Marvicsin and Greenfield 221). The dying machine, barely held aloft in its crippled state, is still capable of murderousness. The trapped pilots may die, but as long as there is some motion, they will use it. Burning fortresses can kill their inhabitants; the helicopter's main rotor is anchored at the top of the aircraft by the so-called jesus nut, and when 'the nut came off, the complete rotor system just ceased to exist in this universe as the myriad component pieces, driven by centrifugal force asserted their individuality all at the very same instant ... [Joe] opened his mike. He screamed all the way down' (Marvicsin and Greenfield 90–1). Cata-strophic machine failure removes any practical agency: screaming is all that is left. Designed for speed, not glide, for full power, not stalls, the helicopter rushes suddenly in the new direction of the Earth at nine metres per second squared, reminding technophiles that, as Chris Gray notes, 'In the war of mechanical speed against human reactions, bod-ies are the only real losers'(Gray 40).

The machine's ghastly qualities are evident to the pilots who airlift the dead. The helicopters are marked by, and mark, their freight: 'Blood was everywhere. It had poured out of the cargo compartment, and the rotor wash from the main rotor blades and tail rotor had effectively painted the tail boom and back half of the helicopter with a sticky coat of blood' (Carlock 208). With the rotor as applicator, the blood is a decoration the machine slaps on, much as the pilots take pleasure in painting signs on the aircraft. Pilots learn not to look back into the crew compartment, not to see the fate of bodies deposited there. In a desperate attempt to clean his helicopter, Mason flies it to a river and hovers in the water while the men wash the machine out. Other pilots describe tilting the helicopter nose down only to have blood rush into the chin bubble, the clear foot wells that allow the pilot to navigate. Blood runs through the machine that appears to consume bodies and certainly disposes of materiel handily, as one pilot recalls: 'One day I flew first-light and last-light missions and every mission inbetween, each about an hour and forty-five minutes long. We refueled and rearmed seven times. Each time we took on 14 rockets, 6,000 rounds of minigun ammo, and 4,000 rounds of door-gun ammo' (Matlock quoted in Brennan 107). The outrageous ordinance expenditure (98 2.75 in. rockets, 42,000 7.62 mm rounds, 28,000 .50 calibre bullets – from one ship in one day) is part of the production model of war prized by McNamara. Like the First World War, the other infamous twentieth-century war of attrition, Vietnam made possible what war analyst Paul Virilio identifies as the 'disap-pearance of men, material, cities, landscapes; and the unbridled con-sumption of munitions, material, manpower' (*Speed* 53).

If humans and materiel can be consumed, so can the delicate craft that carry them. Like wide flying boxes, early Hueys appear to reporter Michael Herr as 'fat poisoned birds' that 'fell out of the sky ... a hundred times a day' (14). Even the unending attention paid to the Hueys by their mechanics and crew chiefs cannot keep wounded machines aloft. Marvicsin passes a downed craft and looks for the shot that winged it: 'It amazed him that a Huey could take a zillion shots and keep on flying. And it also amazed him that just one lucky shot could knock one of those suckers out of the air in two seconds' (Marvicsin and Greenfield 100–1). The fragile tail boom with its few cables and power lines, and the critical small rotor that stabilizes the craft, cannot withstand even the smallest attack (just as removing a kite's tail ends not only the handler's chance of controlling it but the whole kite's ability to fly). A Cav trooper observing a downed Cobra gunship ('known for its ability to bring smoke on the enemy') that has been paralysed by one bullet is reminded 'of my own vulnerability' (Bradley quoted in Brennan 213). The helicopter is not, after all, a flying tank; the human and helicopter bodies are vulnerable to all kinds of miniscule damage. Intended for agility and range, not stability and safety, the helicopter is unable to bear the kind of armour it requires to make it safe and maintain its mobility simultaneously.

The problem of loading protective layers on a relatively frail airframe hasn't prevented recent rotorcraft designers from achieving, writes military historian and enthusiast Edgar Doleman, 'the incorporation of armor and rugged composite materials into the aircraft's skin to resist the impact of small arms fire, the main enemy of helicopters in Vietnam ... The UH-60 [Black Hawk troop carrier] proved its soundness during the American invasion of Grenada in 1983. During the short attack, one UH-60 was struck over twenty times by heavy machine-gun fire ... but the UH-60 successfully completed its mission' (Doleman 166). In fact, the Black Hawk's heavy armour makes it a lumbering truck when compared to the Huey, which appears to be a fleet little race car beside it.

Black Hawks, more heavily armoured than any other helicopters, were downed with some training and little effort by the Somalis during the so-called 'Battle of the Black Sea' (Bowden 110–11, Edwards 47), when in one night 'Two ... high-tech MH-60 Black Hawk helicopters went down in the city, and two more crash-landed back at the base' (Bowden 333). American aircraft manufacturer Sikorsky's Black Hawk, particularly the story of the ship's involvement in the battle, has become a site of trouble for military historians. All agree that things became disastrous when the Rangers, stranded in their mission to kidnap two of Muhammed Farah Aidid's men from his compound, found that their Black Hawks couldn't provide the air mobility on which American troops had so much come to rely. Some say that the troops were under-supported – ultimately it was rolling armoured vehicles that brought out the survivors. Others have argued that the mission was ill-advised and never should have occurred. All agree that the soldiers fought well. Ridley Scott's problem when he came to film Mark Bowden's book *Black Hawk Down* was that he had to choose between telling the story of the men or the machines. In choosing the men, Scott also chose not to deal with any of the politics, with the troops' racism toward the Somalis, with the imperialist actions implied by the raid. He chose not to show Somalis being massacred in a key military event known as battlefield swarming, where superior technology and firepower was overcome by masses of determined, unarmed attackers. While Bowden shows them respect in his account, Scott refuses to present any Somali heroes. Because dead American soldiers were famously dragged behind vehicles through the streets of Mogadishu, the event became an American tragedy, and the three hundred to five hundred Somali dead have been written off. None of these factors affected the continued manufacture or purchase of Sikorsky's Black Hawk helicopters; Sikorsky produces and the Pentagon buys a whole line of heavy-lift helicopters, including the rarely seen Pave Hawk and Pave Low, which themselves dwarf the Black Hawk. Like the Vietnamese before them, the Somalis understood that technology was the centre of American weakness: if the Somalis could bring down one helicopter, they could trap a pilot; if they could trap a pilot, they could lure in a troop of Americans who would come for the live body or corpse. Downed helicopters were the key to American military catastrophe.

Ironically, helicopters are now armed with even more armour-destroying weapons (Hellfire missiles) but themselves remain vulnerable to much smaller munitions, or to the same hand-held optically- or wire-guided missiles that threaten tanks. Rotorcraft have become gun platforms, particularly the Boeing Aerospace AH-64D Apache Longbow, armed with a machine-gun that unloads 625 30 mm rounds a minute, and various combinations of air-to-surface, air-to-air, anti-armour, anti-personnel missiles. Unlike previous helicopters, the Apache can simultaneously fly backwards and upwards, and perform aerobatics that would please Cobra pilot Dennis Marvicsin. It has become as much a war hero of the early twentieth-first century as the Huey was of the mid-twentieth century: Boeing is delighted to report that its product is 'credited with destroying more than 500 tanks plus hundreds of additional armoured personnel carriers, trucks and other vehicles' in the first Gulf War (Boeing 'AH-64'). Boeing and Sikorsky have come together over the last decade to build an entirely new rotorcraft known as the RAH-66 Comanche which, while it can carry missiles and other weapons, is primarily dedicated to speed, lightness, and agility. These corporate actions are aligned with what war theorist Paul Virilio identifies as 'the upkeep of the monopoly [that] demands that every new engine be immediately superseded by a faster one. But the threshold of speed is constantly shrinking, and the faster engine is becoming more and more difficult to conceive of' (*Speed* 46).

The signature thud of the Huey's twin-blade rotor has been replaced by the Comanche's purr, the buzzing susurrus of a wartime pacemaker. Sound files posted on the Internet by Comanche fans are diagnostic of battlespace: instead of heavy pounding, there is now only a feathery whirring – something is there, but it isn't clear what. Even more alarming in its serpentine profile than the Cobra, the Comanche is made of stealth materials and has, according to defence analyst Michael O'Hanlon, 'one-fourth the infrared signature, one-half the acoustic signature, and less than one-hundredth of the radar signature of current helicopters like the Apache' (*Technological* 71). What that means is that, unless you can physically see the craft, you won't find it. It won't identify itself to detection equipment, even high-end technological surveillance. The Comanche's test videos show it blithely dropping for the ground in a ninety-degree bank, executing powered dives, slipping sideways and flying backward at high speeds. Boeing Sikorsky claims that the craft flies in all weathers (a crucial improvement because helicopters were repeatedly grounded in Vietnam by rain or fog), particularly 'battlefield obscurant conditions,' and is fully 'interoperable,' the turnkey phrase for entrance to the mythical realm of C4ISR (Boeing Sikorsky 'Capable').

Yet, for all its wonders, it seems that the dream of the Comanche may be over. In early 2004 Secretary of Defense Donald Rumsfeld snapped off Comanche's rotors, suggesting to defence watchers that the race will go in a different direction. The limits of physical speed will be exceeded by light and information, not mechanical objects like helicopters. The rapidity demanded by the Army in its 1994 training doctrine (United States Army *Force XXI*) and now accepted as integral to the RMA is typical of the information war fought in and through cyberspace. It isn't as if the helicopter will disappear: there's a war on for the war, as the military contests its adherence to the RMA. War and human rights expert Michael Ignatieff looks back to the lessons of Vietnam and warns: 'Generals like Norman Schwarzkopf were skeptical: they had bitter combat experience of both fog and friction in Vietnam. They also knew that the 'systems analysts' of the Pentagon had promised then that new technology married to tactics – the Huey helicopter re-equipped as a gunship – would dispel the fog and grease the friction of war. And they hadn't' (173). The helicopter was an early motion in AirLand Battle, one turn of the wheel in the Revolution in Military Affairs. Whereas Schwarzkopf was dubious about information warfare, others are dedicated to it. Admiral Bill Owens, vice-chair of the Joint Chiefs of Staff from 1994 to 1995, is one of the RMA's strongest advocates. He argues that information technology when connected to rare mechanical creations like the Comanche promise nothing short of omniscience, 'the ability to see a "battlefield" as large as Iraq or Korea – an area 200 miles on a side – with unprecedented fidelity, comprehension, and timeliness; by night or day, in any kind of weather, all the time' (Owens 14). And, while Owens has his critics, people like Rumsfeld and Michael O'Hanlon who see a very different and cheaper way into the RMA all speak the same language of technological dominance.[27]

NASA's report of pilots directing craft by muscle twitches, so that 'machines respond directly to our gestures and our thoughts' (ABC News Internet Ventures), parallels Gray's discovery that 'at Wright-

Patterson Air Force Base ... Human subjects have been trained, using biofeedback signals of flickering lights, to fire brain waves in increasing or decreasing amplitudes. These signals turn that flight simulator left or right, as if they were flying by brain waves' (204). If Rhesus monkeys can drive a robot arm, human brains certainly seem capable of flying craft remotely and hands free. The thought war is the fastest there is, and connecting the war of brain impulses to rotor-craft like the Comanche fulfils the machine logic that entranced the United States a long time ago. The Comanche's powerful muttering began as the Huey's beat that 'imposed a rhythm (otherwise absent) upon the war' in Vietnam (Spark 92). If the Huey 'was the American project of Vietnam ... the ultimate wedding of conventional close combat with the perfection of twentieth-century non nuclear, counter insurgency, rapid-reaction techno war' (Beidler 54), then the Co-manche flown by brain-machine interface promises to abolish battle-space hesitation entirely. The faster the machine gets, the more trans-parent the battlefield becomes, the more the human must be enclosed: human sense can't keep pace with machines. One RAND report cautions that Apache 'Pilots and CPGs [Co-Pilot Gunners] ... would now be encased in a newly designed Kevlar helmet with its liner and a 1-inch television CRT clamped to its side ... It was some-what disquieting for the flight crews to be told that they had 7,000 volts in the cable looped over their shoulder' (Amer and Prouty 14). That's old news. The new helmets feature flat-screen monitors that draw much less power and drastically reduce the threat to the pilot (What if a CRT was hit during combat? What if the pilot crashed with such a helmet on? With flat screens, those worries vanish). Huey visual-display systems use a combiner (integral to a Head Up Display [HUD]) to superimpose targeting information on the battles-cape. The pilot looks through the sight and flies normally. He is free to turn away, or to leave the craft if it is downed or damaged.

The early Reflex sight is, however, useless for conveying other kinds of information. It can't present a thermal picture, can't give the pilot a night-vision display of battlespace, can't legibly display num-bers and characters. There's not enough vision in the sight. And so the once traditional HUD has given way to intricate helmets that have their own built-in sighting system. These so called Helmet Mounted Displays (HMDs) almost entirely enclose the pilot's head

13   **Seeing Straight**: Bell UH-1C 'Huey' Early Head Up Display (HUD), here the pilot's 'Reflex Sight.'

14   **Double vision**: The Kaiser Electronics 'Wide-Eye'™ Helmet Mounted Display has now been delivered to Sikorsky for use in air and rotorcraft.

(fig. 14). 'Off the bore' targeting means the ship doesn't have to be pointed at the target: the helmet scans the pilot's retina to see where he or she is looking, then aligns the guns to match (the guns are a lot smarter than they once were, provide target ranges and adjust for windage). Honeywell developed a Helmet Mounted Display for the Apache helicopter, shifting the emphasis from human to machine: it is the machine that bears the burden – the body has reached its limit. But Honeywell's display was focused on one eye only, and it now competes with visors that wrap around the head. Whether or not the Comanche flies, Kaiser Electronics' binocular helmet is likely to become the standard.

The pilot is now director of a war information-entertainment centre; beside each eye sits a one-inch flat screen, and data pours onto lenses or visors surrounding the face. The pilot is so much a part of the machine's furniture that ejection, requiring the severing an umbilical full of data and high voltage, is next to impossible (Amer and Prouty 14). Flesh vanishes as technology gradually encloses it. The incarcerated human rides a raft of machinery on a digital flood, relying on technology to impose grids upon, make manageable and interoperable, the information glut. The push to speed is so urgent and all encompassing that it's likely fewer humans will occupy battlespace. It's only been a few years since the dream of air mobility in Vietnam, but, for computing power and speed, those years are generations. Humans still see the same way no matter how much video games have retrained them, but machine vision sharpens, clarifies, and becomes more penetrating at each surge in micro-processing. The early trip signified by the steady heart thudding of the Huey rotor blades, now replaced by the Comanche's feathery whirr, is giving way to a new world of battlespace dominance that will be, if we accept RMA doctrine, completely uninhabited by humans. And yet it will be full of other kinds of war-directed life.

Chapter Five

Dead Slow:
Loitering in Battlespace

It's more than meets the eye! First off, when we see that nice airframe, we have to remember that behind that airframe is a wonderful command and control system. Links, systems, integration – all the things that are really hallmarks of effective systems today.

– Major General Paul Nielsen, commander, Air Force Research Laboratory,
Wright Patterson AFB, at the 27 September 2000 roll-out of the Boeing
Unmanned Combat Aerial Vehicle X-45A

A screaming comes across the sky. It has happened before.

– Thomas Pynchon, *Gravity's Rainbow*

Airless up here. Lively though. Vehicles, propeller and rotor driven, pass each other with a vague motor susurrus, a droning buzz (a reverberant word for blitz survivors). AirLand Battle day and night – there's a flying Internet circus aloft, each UAV is its own node in the worldwide war web. Data packets surge between airborne servers, low-flying spy eyes, high-altitude satellites, and loitering hunter-killers: no operators need direct the information traffic. The aircraft engines, sometimes even the rockets, are air-breathing, but there's no inspiration, no lungs at work; it's all breathless soaring, motion without detection, presence without knowledge, sensation without feeling, dead things aloft in the clarifying medium of battlespace. Near the ground are little movers, disks with fins, tiny flying wings, pins that flap, objects that confused birds flock to, fly with, or try to eat. Some craft dwell a while, some make busy little circles over things that intrigue them, some fly into rooms or behind city walls, perch, stare, make contact with other craft still in motion: information is on the move. Between the lowest uninhabited aircraft in the radar grass, at 15 or 60 metres, and the highest, turning in lazy endless circles days at a time, running on sun at 30,000 metres (29 kilometres), empty rotorcraft orbit the territory, taking orders from their onboard intelligent agents only. In the middle of the stratosphere where oxygen is rare, scoops pick up what there is and ram it into engine intakes – the mixture must be rich enough. Getting hot, the vehicles need to dump extra energy and stay cool (they're always calm) – physical laws are the strictest ones in this particular battle. The self-managing web communicates with humans somewhere, maybe on the ground, maybe in the air, maybe on this continent, or another. Decisions made, the little watchers pull back and the stand-off weapons, which have been hovering with 'deadly persistence,' roll in. Larger craft are followed by even larger, until the combat weapons arrive – loiterers cede the battlespace. The weapons, too, are uninhabited but still see and think. Enforcer craft bank over the targets at speeds and in manoeuvres that would kill humans, bombarding the ground with microwave pulses, shots of energy designed to kill other machines or 'actively deny' the ground to human bodies. If there's screaming down there, no one in the air can hear it because there's no one in the air.

Key to interoperability and operation in battlespace is the idea that commands must share data, not just materiel and forces. Recent focus has been on the Joint Surveillance and Target Attack Radar System and the Lockheed Martin F-35 Joint Strike Fighter (JSF). The JSTARS flies high and looks down, providing all ground forces with continuously updated information. Behind the JSF is the idea that all American and international forces will buy the fighter, allowing the battlespace commander to manage the whole system as an interoperable network of aircraft. However, in the air, on the ground, or at sea, the intent is that forces be joint at national and international levels. Even if countries can't afford the JSF, they can lay out the necessary money for a few UAVs, which are cheap and, if necessary, disposable. UAVs are an information technology teleoperated electric blanket that keeps the Revolution in Military Affairs warmed up. They range in size from big insects to craft with wingspans longer than those of Boeing 747 jets. Most UAVs can taxi and take off from standard runways, be launched from ship decks, or be thrown into the air by catapult or Jet Assisted Take Off (JATO). The current United States government is buying UAVs at a pace that outstrips their acquisition by the highest-tech defence administrations in U.S. history (those of Robert McNamara [1961–8] and Caspar Weinberger [1982–7]). Donald Rumsfeld's Department of Defense has forecast a five-year $7.1-billion, which by 2010 will top $10-billion, investment in UAVs (Office *Roadmap* 19). The administration's dedication to UAVs is in line with RMA policies of investing in high technology that, networked and interoperable, produce a transparent battlespace. UAVs aren't designed to operate on their own but in conjunction with other such machines: the connections will be made over wireless Internets and satellite links. The RMA vision sees UAVs collecting battlespace data that is accessible to all the forces. The UAVs themselves should require a minimum of technical expertise, making them adaptable to the entire military. On their own, UAVs don't yet carry enough firepower to fight an air war, but, in conjunction with human air forces, UAVs are what the military adores: force multipliers, agents that greatly magnify combined human and machine strength. The force multiplier alters the battlefield power ratio in a way that is catastrophic for the enemy; putting a saddle on a horse meant that a rider could more easily control a horse, but the addition of the stirrup made the horse soldier able to fight with two hands while carefully guiding the horse by foot. The horse was important before the addition of the stirrup, but afterwards its power was magnified geometrically. One horse and one rider with two stirrups multiplied, rather than simply adding to, the warrior's force. The UAV is perceived to be the military's new stirrup. UAVs are part of the military's desire for what it calls 'endless air occupation': the ability to stay in the air indefinitely and with impunity (Carmichael viii).

Nazi General Walter Dornberger and Werner von Braun's epiphanies about robotic aircraft at Peenemünde, when made into weapons by Nordhausen concentration camp inmates' slave labour, transformed the First World War glide bombs into the *Vergeltungswaffen-1* and 2 ('vengeance weapons'). The V-1 and V-2 respectively became today's cruise and intercontinental ballistic missiles. But UAVs are defined as returnable, reusable aircraft, where the remote operator is elsewhere in the air or on the ground: some UAVs fly unattended, guided by their onboard computers. Their single-mission nature excludes glide bombs, artillery shells, and ballistic or cruise missiles from the UAV family. Both the V-1 ('buzz bomb' or 'doodlebug'[28]) and the V-2 were one-way weapons designed to fly over long ranges and deliver explosives that incidentally resulted in the weapons' destruction. *Vergeltungswaffen* technology is at the core of the uninhabited vehicle: the buzz bomb could be programmed to fly a given distance (its air log recorded mileage) and stay level on course at a given height (gyroscopes, compasses, and barometres handled these tasks). When the air log matched the pre-set distance, the motor shut off, the buzzing noise stopped, and the bomb fell, a terrible silence signifying the target. Like many UAVs, the V-1 was automatic, programmable, and autonomous: a microfilament divides the V-1 from early UAVs.[29] Between 1960 and 1972, the most famous of America's early UAVs were fashioned from target drones originally designed to provide tests for Air Force pilots and Navy gunners. The promotion of autonomous jets from targets to Special Purpose Aircraft was caused by the 1960 downing of Francis Gary Powers and his U-2 spy plane over the Soviet Union. Dedicated to the 'mutual surveillance' between the Soviet Union and the United States that the Eisenhower administration claimed would prevent further missile madness, the Ryan Aerospace Firebee BQM-34 target drone was fitted with a camera, tested, and then flown over China, the Soviet Union, Cuba, and, finally, Vietnam. Despite their drawbacks (the film canister, called a 'scorer,' often had to be recovered under dangerous conditions, and the craft were initially difficult to control), Firebee drones flew thousands of mission hours collecting data considered indispensable by the Air Force. The turbine-powered swept-wing Firebee bears an uneasy resemblance to the buzz bomb: it was the shift in mission from explosive delivery to information recovery (a sign of post-industrial, information war) that distinguished the machines.

UAVs fly for varying amounts of time, from a few minutes to a number of days, require differing amounts of instruction from the ground (there are ten basic levels of UAV autonomy), and carry payloads such as surveillance gear, smaller uninhabited craft, electronic markers, bombs, and energy weapons. While the American war in Vietnam confirmed UAVs as important to the defence establishment's war kit, watching Israel use UAVs to direct the vivisection of larger and better equipped enemies convinced the United States to invest fully in drones. UAVs can by classified by the distance they can fly, the height they can attain, and the amount of onboard computing power they have. The most powerful of the family is Lockheed Martin's long-flying (with one intriguing exception), far-reaching, intelligent Global Hawk, defined as a HALE UAV (high altitude, long endurance). Next down in size and range is General Atomics' Predator, of which the latest (the Predator B and Predator ER [Extended Range]) are armed with a variety of weapons. Then follow a series of craft that can take off from runways, aircraft carriers, or battleship decks, fly for roughly eight hours at a moderate height (4,500 metres or about 5 kilometres), and whose main job is to return information to their users. These can be grouped together as HSM (high speed, manoeuvrable) craft. Below them is a class of craft that can be unpacked from two large suitcases, assembled in less than ten minutes, thrown into the air, and controlled by an operator with a laptop or virtual-reality helmet. Smaller and lower flying, with less duration (usually called 'loiter' or 'dwell' time), are a series of micro- or mini-UAVs that are put into slingshots and fired into the air. At the bottom of the range are tiny UAVs modelled on insects. All of these come under the third UAV classification: disposable. UAVs, whether they are kept aloft by solid or air-breathing rocket, turbine, pusher-propeller, rotor, chemical or solar power, are expected to accomplish one or all of the three 'D' jobs: the dull (long-range reconnaissance), the dirty (sample clouds of NBC materials), or the dangerous (missions in which the sacrifice of a human life would either be unwarranted, too costly, or embarrassing, as the Gary Powers U-2 incident was for the Eisenhower administration).

Because Revolution in Military Affairs warfare is based on information that arrives from diverse sources, and because that information must be accessible to all fielded units, it makes sense that the RMA would require a movable Internet. Dropped from aircraft, thrown into the air by hand, slingshot, catapult, or jet, UAVs muttering into battlespace on real-time surveillance missions interconnect to form a network of networks that produces the RMA's required metadata. Only a few UAVs do the work (Global Hawk, Predator, Pioneer, Shadow 200, Pointer, Dragon Eye, Desert Hawk), although some forty companies make 140 craft worldwide. Most UAVs are equipped with advanced optical systems, synthetic aperture radar, high-speed modems, or satellite links. All transmit real-time video (some in colour, some in High Definition Television [HDTV]). Since American attacks on Afghanistan in 2001, the Predator A has been, in the defence establishment's plastic language, 'weaponized.' The 2003 war on Iraq accelerated the UAV's transition from information agent to hunter-killer. When DARPA commissioned Boeing to create a swarm of hunter-killer UCAVs and UCARs, the next turn in the RMA occurred. The UCAV (now a generic term – in 2000 UCAV referred to Boeing's X-45A UCAV) will 'perform combat missions that do not currently exist; high-risk missions that do not warrant the risk to human life; or current missions that UCAVs can perform more cost effectively than current platforms' (National 17). The cluster of descriptors is revealing: new machines make new warfare possible. By its very existence the UCAV will create a kind of combat typified by hard against soft, machine against organism. The UCAV will outfly and eventually outgun vehicles driven by biological packages (humans). Humans will become assisted devices, organisms that require life supports like air and water. And then there's the cost. The U.S. Air Force wants to buy as many of the F-35 Joint Strike Fighter (a fully interoperable fighter-bomber) as it can. The F-35 costs about $30–35 million USD. Boeing's X-45C UCAV, Northrop Grumman's X-47 UCAV-N Pegasus, and RQ-8A FireScout UCAR will each cost about a third of the JSF.

Since the machine may be lost but the UAV crew will be on the ground, hidden by walls, hardened bunkers, or distance (the controllers may operate the craft from another continent), there will be no widow-making and no embarrassing prisoners of war. The UCAV will be put 'out front, in harm's way,' as the Defense Department's Joint

Robotics Program motto has it: 'UCAVs are foreseen as a 'first day of the war' force enablers that complement a strike package by performing either preemptive lethal destruction of sophisticated enemy IADS in advance of the strike package or provide supportive Electronic Attack (EA) against specific threat emitters' (Office *Roadmap* 178). The package in question consists of actual people flying aircraft. Further decryption follows: UCAVs will be used before the war has really begun or in the first-wave assault; they will crank up more conventional forces' power by cleaning out and seeing through battlespace; they will attack the enemy's sensory organs (IADS, the Integrated Air Defense System, otherwise known as the eyes and ears of ground installations that destroy attackers), as well as take down transmitters and missile batteries that pose threats to friendly forces. UCAVs are extraordinarily sleek flying wings, deltas without tails, craft that are almost impossible to see and are invisible to radar. Their airframes are presumably made of stealthy non-metallic materials like carbon composites or Kevlar, honey-combed structures that diffuse radar beams, and are topped off with radar-absorptive paint (even the fuselage joins are kept miniscule so that they need not be covered with stealth tape, as are the joins on B-1 and B-2 bombers). They will be the force before the force, a concept already outlined by mobile-war theorist Heinz Guderian. The difference is that, instead of linking air and land power, here it is air and air power that are connected; one air blitz precedes another. The UCAV will, in every fashion, make the way clear for humans. Empty battlespace scrubbed clean of jamming devices, radar and infrared sighting systems that give anti-aircraft guns their targets, phone and data lines that carry warnings, will be occupied by perceptive uninhabited craft.

UAVs are being ordered in quantity because of a 'combination of "mission pull" (e.g., risk avoidance and cost avoidance), which require that systems be developed for certain missions, and "technology push," which is fueled by advances in particular technology (e.g., microelectromechanical systems [MEMS], electronics, and composites)' (National *Uninhabited* 22). The decrease in machine size, increase in computer power and storage (UAVs will get brighter), and the RMA's fetishization of technology guarantees that post-human warfare will be autonomous, sophisticated, and invisible.

RMA theory insists on clear battlespace. Hundreds of U.S. military Powerpoint briefing slides show transparent domes, like 1950s diner cake covers, enclosing large geographic regions, where jagged lines (signifying electronic communication) connect the military baked goods inside the cake display. Maintaining clarity means not only purging battlespace of the enemy's communications but also preserving your own network's invisibility. Just as soldiers will wear armour that erases their heat signatures, and angular radar-proof tanks will slouch lower on the field while honey-combed stealth rotorcraft hover overhead, so must UAVs that aren't already cloaked for low visibility (they look like flying slices and saucers, thin deltas that leave no vapour trails) learn to disappear. The simplest way to provide stealth is to make UAVs so small they can't be seen. Advances in micromachining have brought tiny UAVs within reach. Etched like computer chips, MEMS machine parts a few millimetres square can be mass-produced. The MEMS industry has promised DARPA the ability to establish what it calls 'systems-on-a-chip,' a miniaturized version of the RMA's system of systems (MEMS Exchange). In 1996 DARPA opened a design competition for micro-UAVs (sometimes called MAVs), with the arbitrary maximum craft size of a six inch square. AeroVironment's Black Widow quickly took the lead: 'The Black Widow is a 6-inch span, fixed-wing aircraft with a color video camera that downlinks live video to the pilot. It flies at 30 mph, with an endurance of 30 minutes, and a maximum communications range of 2 km. The vehicle has an autopilot, which features altitude hold, airspeed hold, heading hold, and yaw damping. The electronic subsystems are among the smallest and lightest in the world' (Grasmeyer and Keennon 1). The diminutive Black Widow (it's about forty grams, the weight of a soup spoon) is significant not only for battlefield invisibility but as a sign of UAVs' fragile strength. The remote pilot lands the Black Widow by dropping it at a shallow angle onto the ground. The craft has been damaged more by its operators as they handle it than it has by flying or landing (trouble caused by touching the UAV has been resolved by shooting it from a small box launcher). The Black Widow's vulnerability to human hands is diagnostic of the difficulties UAVs have connecting to people: uninhabited vehicles require uninhabited methods of delivery and return – humans make trouble for vehicles. The Black Widow and AeroVironment's next generation of micro-vehicles (the Wasp and Hornet) are intended for ground units fighting Military Operations in Urban Terrain. Micro-UAVs can fly over buildings, into rooms, see who's there, what weapons they do or don't have, and, if the UAVs get shot down, knocked down, or hit walls, they can be written off because of their relative cheapness. The Black Widow runs on expensive short-lived lithium batteries, the Hornet on solar power. Neither are desirable power solutions, so micro-UAV producers have looked increasingly to miniature gas turbines (centimetre- or even millimetre-diameter generators – 'button size' appropriately captures their scale [National 77]). For more than a hundred years, humans have worked collectively on the gas-turbine engineering project that has absorbed an estimated $50–100 billion USD worldwide, and MEMS machines will make button-size turbines possible in the next decade. Those turbines will drive micro-UAVs.

Speed and cheapness are the micro-UAV's kernel: taking chances on new technology means risk, slow production, and expense. Much of the UAV world relies on what are blithely called COTS (Commercial Off the Shelf parts and technology). However, one particular UAV uses two unusual new technologies. The entomopter, more affectionately known by its creators as 'Robofly' (after the 1987 film *Robo-Cop*), is one of the few surviving projects that relies on flapping wings to keep it aloft. Even more unusual is its power source, a chemical cocktail the operator injects into the craft's motor. The chemically powered flapping wings returns power to the onboard chips that control the craft's flight – it's called Reciprocating Chemical Muscles (RCM). The entomopter's wing-stroke is much simpler than that of a fruit fly, whose flight pattern is a gorgeous lacework of whorls and arabesques. Micromechanical flyers like Robofly are designed in keeping with MIT roboticist Rodney Brooks's idea that artificially intelligent robots should be 'fast, cheap and out of control' (the main difference is that the Robofly will be tightly controlled). The insect model for the entomopter indicates how numerous, persistent, and disposable the micro-UAVs are to be. Etched in batches like silicon chips, Robofly would appear on the battlefield in swarms. Entomopter models the public has been allowed to see are quite large (seventy-five or more

15   **Seeing Little**: Micromechanical Flying Insects (MFI), one of the official names for Robofly. Although DARPA claims to have stopped funding this project, other sources indicate that it continues.

millimetres in length): the final designs call for ten- to twenty-five-millimetre stainless-steel microflies weighing forty-three milligrams (about the weight of a raisin) (Bonsor).

The whole entomopter process is one of pain: the insect is a phobic site for many humans, and the chemical injection into the insect's thorax is equally discomfiting to see. Pain, phobias, surveillance, amok chemistry, mechanical swarms: the entomopter looks like a marvellous little nightmare, and so is attractive to developers. It's difficult to tell how advanced the entomopter has become since the current picture of insectile aerial robots is like the UAV industry: dark and indistinct, where many projects are 'black' or 'closely held' (deeply classified). In 2001 it seemed that DARPA had written off flapping-wing, RCM flyers. The UAV industry and governmental agencies are focused on silence and refusal to answer. A recent discussion of micro-UAVs between *Aviation Week & Space Technology*, trade journal for civil and military air industries, and British Aerospace (BAE) had this result: 'While BAE Systems personnel would not address the concept, U.S. Air Force officials indicated that the service has been looking at small UAVs that could fly within a few feet of an antenna, possibly even attaching itself, and then disable the site with an energy spike, jam the signal or insert false targets. By being that close, the jammer would require only relatively low power to achieve great effectiveness' (Fulghum and Wall 'Small' 60). Some form of microflyer will have to grapple with and spike the antenna, but the people who make the object wouldn't talk, and only the government threw *Aviation Week* a bone or two. The entomopter disappears from view not only because of its size but because information about it has been made to vapourize. The systems are about stealth, and that means the concepts have low-information cross-sections: they're off our knowledge radar.

Disappearance has always been part of the UAV world. In the early days of drone aircraft, Ryan Aerospace learned to turn a blank stare on all questioners: 'If they [the Communist bloc] shoot one down and announce it publicly, don't deny it; but don't acknowledge it. Just reply, "no comment," and sweat it out' (Wagner 57). The dictum seems to be that, if the machine doesn't reveal itself, the engineers shouldn't either. Operations have only blackened since then. News that there *could* be a small craft with feet sticky enough to grip transmission equipment and fry it into silence (jam), or better yet, mislead the equipment into thinking that it is another kind of craft (spoof) is unsurprising: UAVs fly in the interoperable battlespace of could, might, and possibly. The value of bad information, to either the public or the enemy, soon became clear to the Firebee engineers: 'The idea behind the D [a mid-size Firebee] was to equip the drone with a traveling wave tube (TWT) to make it look like a much bigger U.S. aircraft. The drone would then be flown over Cuba at night or just before dawn to draw SAM [surface-to-air missile] fire; to bring a SAM up where the CIA electronics gear could effectively get the fusing and beacon information from the Soviet SA-2 missile' (Wagner 46). The drones were already mock aircraft, and their users learned to put systems into systems, to create meta-drones. In this instance of spoofing, the drone lures out a surface-to-air missile that, once locked on, reveals the missile's base. Then the base can be bombed. Bad information brings forth good information, an integral part of UAV operations.

'Jamming' and 'spoofing' are terms marked by the countercultural cybersizzle of information warfare. But UAVs fit into a much older, Manichean, vision of the world. UAVs are designed to stay in the dark and bring the enemy into the light where they can be destroyed: they are formed for seeing and not being seen; are organs of tempta- tion (teasing out the enemy's weaknesses) and omniscience; are fitted to have knowledge and deny it to others. In an irony that under- pins the world of black UAV programs, 'goodness' or 'lightness' (useful information about the enemy) is expected to emerge from darkness. The diagnostically named Black Widow (it has a black and white hourglass on its top surface) turned out to be capable of spoof- ing the organic and machine worlds equally; its developers noted that they had 'seen sparrows and seagulls flocking around the MAV [Micro Air Vehicle] several times.' Their further comment that the Black Widow 'looks more like a bird than an airplane' (Grasmeyer and Keennon 8) is offered without ironic references to Siegel and Schuster's Superman, without consideration that the craft is Super- man's cultural antithesis. A slippery membrane divides the natural from the human creation. UAVs are alive with discomfort for humans. In technological jargon the Black Widow is what is called a 'testbed' for other craft; AeroVironment's subsequent Hornet and Wasp can carry cameras and transmit real-time video to ground con- trollers. In the spring of 2003 there was enough discussion about arming these flying wings with small calibre weapons to suggest that the reality was close by. Arming UAVs brings them closer to the light, requires an increase in their size, but, most of all, moves them from a world of seeing to one of seeing and killing.

The January 2003 *Aviation Week Aerospace Source Book* ran, near the index to UAV manufacturers, a General Atomics advertisement for the Predator B, the UAV industry's current centrepiece (fig. 16). Against the backdrop of the Earth seen from space (perhaps the moon), the Predator leans into a turn and looses a Hellfire missile from under its wing. Sharp against the dark ground is the Predator's

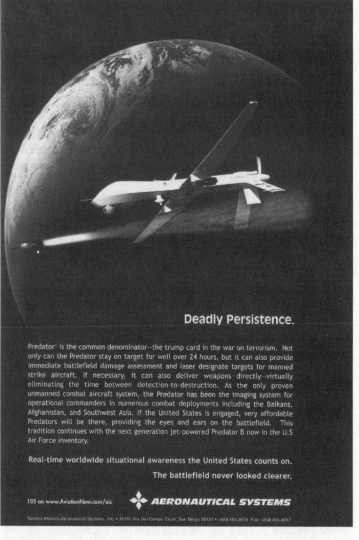

16   **The World Is My Predator**: General Atomics touts the clearness of the battlefield and the stability of its battlespace platform, an odd duck-like craft with its diagnostic bulbous nose containing the avionics package.

chin bubble, a pod stuffed with optical sensors and cameras. According to the graphics, the Predator commands the globe, sees into the dark Earth from an invisible, lofty, vantage point. The Predator can't fly as high as the ad (if taken literally) indicates, nor would it have a target for its missile at that altitude. The ad is focused on global omniscience and uses darkness to limn the Predator's newfound capability: vision coupled with killing.

In a typical synergistic military-industrial defence motion, the General Atomics' ad tagline, 'Deadly Persistence,' was picked up by the military press that elevated the phrase to the doctrinal level (the concept has now shuttled back and forth so energetically that it's impossible to tell just who originated it). Just before the second Gulf War was declared officially over, Air Force officials told *Aviation Week* that 'the star of the war ... was a "ruthless, staring constellation looking at Baghdad" made up of UAVs, U-25 and other intelligence gathering aircraft that blanketed Iraq for weeks before the actual fighting started' (Fulghum 'New Bag' 22). The Predator and the globe, deadly persistence, ruthless staring: three signifiers with one signified – seeing is killing. The Predator was flagged with its 'deadly persistence' label because of its successes: all the service arms fought for video time, requiring the Air Force to keep its machines in the air as long as they would fly. The 'ruthless, staring constellation,' a network of aerial 'stars' (as one *Aviation Week* writer calls them), suggests there is a new firmament of power in the night sky. The enemy is blanketed not by bombs (although they may follow) but by vision. Battles in the second Gulf war moved relatively quickly because the war started before it started. Before the overt war, the landscape, the installations, the materiel, were stared at. When it came time to bomb or roll in under armour, enemy transmission stations, communication nets, and guidance systems were already offline, and UAVs were elsewhere performing new tasks.

The popularity of Predator A (primarily a surveillance pusher-propeller craft), Predator B (the recently minted, armed jet-powered model), and Predator ER (the extended-range model carries additional fuel tanks) tends to eclipse the high-flying, long-range, long-loitering serpentine jet-powered Global Hawk, which mapped Iraqi battlespace for the military. The Global Hawk 'flew 3% of all aircraft

imagery-collection sorties and only 5% of all high-altitude reconnaissance missions, but it collected information on 55% of the air
defense-related time-sensitive targets. Global Hawk's score card
included locating more than 13 full surface-to-air missile batteries,
50 SAM launchers, 300 canisters and 70 missile transporters. The
UAV also found 300 tanks, which amounted to about 38% of Iraq's
known armor' (Wall and Fulghum 'Intel' 63). The passage's rhetoric
does everything but stand up and cheer at these numbers, reflecting
the voice of combined fanatic baseball parent and statistician: 3 per
cent vision brought in 55 per cent of the intelligence. (Ghostly excla-
mation points hover between the awestruck sentences.) Global Hawk
brought back a data miner's main lode. It saw more per surveillance
foot than any human operator could or did. No wonder that the Air
Force, not generally given to poetic phrasings, calls this kind of performance 'ruthless staring': it well captures the abilities of a vehicle
that can fly more than twenty-four hours at a crack and return such a
density of information. The Firebees that had flown so long before
(really only about twenty-five years) were hard to control, used 'wet
film' (analog – as opposed to digital – systems produce images that
must be retrieved physically instead of being transmitted by modem),
took pictures only of what they flew over directly, were invisible to
the operator stuck on a nearby waddling DC-3, and, all in all, seemed
like they came from another universe. The Global Hawk (as with the
Predator and most mid-size UAVs) taxis and takes off from runways
and needn't rely on the Firebee's arcane mid-air parachute and helicopter recovery system worthy of Heath Robinson or Rube Gold-
berg.[30] The Firebees flew thousands of collective hours and brought
back a great deal of useful information, but they looked at the ground
dumbly, flew blindly, and had to be retrieved mutely (if they could be
retrieved at all); they saw little, and so could kill little.

In the 1990s the United States Air Force and the now-dismantled
Defense Airborne Reconnaissance Office looked to UAVs for further
signal intelligence (SIGINT), electronic intelligence (ELINT), measurement and signature intelligence (MASINT), jamming, surveillance, and aid to ground troops needing to see 'over the horizon.' But
they didn't expect to arm UAVs. At the end of the Vietnam War,
Ryan engineers, having tested the Firebee through thousands of com-

bat flight hours, knew that they could successfully arm their drones. In late 1971 a customized Firebee with an onboard (if low-resolution) black-and-white television camera was tele-piloted near its destination and then commanded to launch the Maverick air-to-surface missile under its wing. Because the Maverick is a smart munition with either an electro-optical or infrared imaging system in its nose, the pilot was able to guide the drone's missile to 'a direct hit [as if] riding astride the speeding weapon' (Wagner 182). It seemed like a *Dr. Strangelove* metadrone, with one remotely piloted craft flying another. The Vietnam War's slow grind to a halt meant a parallel UAV stall, and armed uninhabited craft were set aside. Then, Israeli battles with Syrian forces in the early 1980s reminded the international defence community that UAVs could reverse war fortunes. In RMA parlance, the Israeli Defense Forces denied Syria air space and ground attack by destroying almost all the Syrian surface-to-air missile launchers. Targeting information came from apparently insignificant and hard to pinpoint Israeli Pioneer drones buzzing around the Bekaa Valley. Impressed by the UAVs, the United States formed the joint Pioneer UAV Corporation (uniting Israel Aircraft Industries with AAI Corporation).

UAVs proved so useful in the 1991 Gulf War that the Defense Department determined to acquire wholly American UAVs with greater range and capability than the Pioneer. The Predator A, an unlikely looking craft with a beaky nose, ungainly wings, and strange duck tail, has turned out to be murderous. It flies at 7,500 metres, almost twice the height of the Pioneer, carries four times the payload, and can loiter 'on station' for a day. While the Pioneer has a nose full of cameras and sensing equipment, the Predator can send back high-definition real-time colour video to operators watching non-interlaced screens; the receivers can then redirect the stream to aircraft or ground troops.[31] Where the Predator A is consumed by surveillance, the gap between seeing and killing has been narrowed recently, illustrated by two famous examples. In May 2002 a reporter was shown an aerial video of an Israeli neighborhood bombed by a Katyusha rocket; the narrative concludes this way: 'Moments later a motorcycle darted from the grove, entered the local highway and apparently made a clean getaway. The film continued as the motorcycle eventually stopped and its rider entered his house. The house then exploded as the Israel Air Force retaliated with an air-launched missile' (Fulghum and Wall 'UAVs Validated' 26). The retaliatory strike is a perfect showpiece: the original attack is an ambush, not a stand-up fight; the neighborhood the attacker flees to is a peaceful place – that means it's civilian; and the counter-strike is so precise that only one person is targeted (the house could be full of innocent non-combatants, women, children, and so on, but it isn't a perfect world). Above it all, quite literally, has been a loitering UAV, on station seeking just such attacks. The UAV's vision banishes the idea of escape: panoptic aerial sight lets the Israeli Air Force drop out of nowhere. The idea of 'deadly persistence' and 'ruthless staring' is not so much to produce bodies as to deter through endless surveillance.

The United States had similar success in Afghanistan when it linked an unarmed Predator and an AC-130U Spectre gunship. The AC-130U is the latest iteration of the AC-130 gunships. It is a weighty, four-propeller side-firing craft that must turn and tilt its gun-side at the target; among other weapons it carries a 25 mm gatling that, thanks to twin electric motors, sprays out 18,000 rounds a minute; there's a 40 mm Bofors cannon thunking out shells a little smaller than beer bottles at a hundred shots a minute; and the big gun is a M102 105 mm cannon (the original calibre for Abrams tank gun tubes before they were 'upgunned' to 120 mm), which fires between six to ten rounds per minute. At full operation, Spectre literally pours metal casings out its doors, with tracers from the ammunition belts drawing a straight line of light at the ground.

In the fall of 2001 the Predator began loitering in Afghanistan's airspace. Seeing an RMA opportunity, Air Force chief General John Jumper declassified an eight-minute high-resolution video showing a ground installation under combined surveillance and attack from a Predator and AC-130U Spectre gunship. The viewer hears a number of different voices: the Predator controller, the ground commander, the AC-130 gunship crew (as well as the dense throb of the aircraft's engines), and other control crew in the background. The slow, orbiting point of view we see from comes from the Predator: that same image feed appears on ground command and the AC-130 cockpit's screens. Having warned the AC-130 crew to avoid the 'large rectangular building' (identified as a mosque), the commander begins calling in targets and the AC-130 105 mm cannon fires, the crew becoming audibly more excited as strike upon strike destroys the area. With one building in ruins, a group of trucks engulfed in secondary explosions so bright they overload the Predator's optics, the gunship is directed to follow the Predator's eye, which sees individual humans running across roads and into nearby fields. They are then obliterated by high-explosive rounds. One person's luck holds out until the last explosion envelops them: the Predator sees over the berm that ordinarily might have provided cover. The Predator loiters throughout, keeping the enemy in the light. Jumper told his Air Force audience to whom he first showed the video: 'This was not done six months ago. It is done every day today [*sic*] and it is getting better by the day' (McIntyre 'Rare View'). The video makes the viewer complicit in the kills – after a short time, the viewer's mind eliminates the lag between what is seen and what is next destroyed – and shows the Predator-Spectre union as being unbeatable: all that is loved by the camera is penetrated by the 105 mm cannon. The excitement of standing over a barrel and shooting the fish in it is underscored by the increasingly hysterical voice of the man (probably the target spotter) who becomes so wrapped up in the mission that he starts to yell his 'get ready' commands the way a child screams 'boom' while playing with a toy cannon. As long as airspace is uncontested, then the Air Force can take all the time it wants to look and, having looked, to kill. The film ends with a further demonstration of the uselessness of hiding. As one runner heads for a tunnel, both craft swing around, the Predator focuses, and a series of explosive flowers collapse two tunnel mouths. One Air Force official commented: 'There is a beauty in that video ... Those that listened heard a Predator pilot who was many miles away talking to the fire control officer in the back [of an AC-130] with both of them looking at the streaming video and very, very carefully pointing out which [buildings] were targets, which were mosques' (Fulghum 'UAVs Whet' 53). After all, the RMA's promise is not to stop killing but to kill the right people based on the right data, and to keep friendly soldiers out of harm's way. The film's release is a sign to the military and, secondarily, civilian world, that the Revolution (in Military Affairs) has been televised. As the one-way attack ends, all that is left standing is the mosque: it is unclear if any people are left alive.

The success of the 1972 Maverick-laden Firebee meta-drone is probably responsible for the slow development of armed UAVs: the Firebee special not only removed the pilot from the cockpit but also took his finger off the trigger. But if the Air Force didn't want sacrilegious armed drones buzzing around, that didn't stop DARPA from issuing a call for hunter-killer UAVs, which have now been flying in American airspace for at least three years. The UCAVs are stealthy, and stealth is good, but size is better: for UCAVs to remain relatively cheap, they need to stay small, which means that they cannot muster the kind of obese firepower as can the AC-130 Spectre airlifter on steroids. The drive is on for new, smaller, more destructive weapons for UCAVs. Four different weapons are being pushed by DARPA for UCAVs: special small bombs; an odd slug-throwing device; particularly grim incendiaries; and directed microwaves.

Small Diameter Bombs (SDBs, sometimes SSBs, 'Small Smart Bombs') are the first stop for UCAV buyers. SDBs weigh about 100 kg, hold 20 kg of explosive, and have the same destructive force as 800 kg bombs designed to attack hardened bunkers. Next, the Navy has invested in a black project that has been tested, but otherwise remains invisible, called a 'flying plate' weapon, consisting of an exploding canister that fires dozens of copper disks perpendicularly at a target. The disks, backed by a combination of rubber and explosive, strike so-called 'soft targets' (buildings, houses, humans) at high velocity, either tearing the structure (and bodies) to pieces or coring the material, depending on the way the disk fuzes have been set. The flying plate weapon is like a gargantuan explosive shotgun blast. Third up is another bomb – the intermetallic incendiary. Another black project developed by the Naval Surface Weapons Center, the intermetallic bomb produces a burning titanium-boron cloud that surpasses firestorm heats at 3,700°C (6,700°F). The only thing that stops an intermetallic incendiary fire is the lack of combustible material. Pouring water on a titanium-boron fire generates a secondary reaction that creates even greater heat (USAF Scientific *SAF/PA*). If used against hardened bunkers, the bomb would suck the oxygen out of the shelter and suffocate and cook the inhabitants, blowing out their eyes and lungs with the blast over-pressure usually associated with Fuel Air Explosives. The Naval Surface Weapons Center claims that such weapons are designed to destroy chemical and biological weapons from the air, incinerating all lethal agents. But once a bomb has been made and slung under a UCAV pylon or fixed in its payload bay, the designer's plans are irrelevant.

The final weapon currently favoured for the UCAV goes by a number of names: Directed Energy Weapons (DEWs), Active Denial Weapons (ADWs), and High-Powered Microwave (HPM) weapons (the present abbreviation choice). No matter what it's called, a gun is a gun; the UCAV carries a focusing beam, targeting system, and Internet technology needed to communicate with the operators. HPM weapons fire in very short (millisecond), tightly grouped bursts; the near-future UCAV loaded with an HPM weapon will be able to fire at '100 targets with 1,000 pulses of energy in a single sortie' (Fulghum 'USAF Acknowledges' 27). Air Force tests have shown HPMs wiping out all but the most hardened ground-computer networks, which are scrambled, crashed, rebooted, or destroyed outright by beam attacks. The term 'hardened' has a machismo about it that the military clearly likes (those 'long-rod penetrators' exist, after all, to destroy 'hardened' bunkers), but in this case the 'hardening' is electronic – networks able to withstand energy weapons must be completely self-contained, closed into hermetic vaults with no lines to the outside world. In an energy-weapon attack, the power fails, radar stops tracking, and enemy air defences are scuttled. While the UCAV is set to 'fly against' ground installations and knock out materiel, computers, networks, and information sources (called 'chip-frying' missions), there have been musings about microwave weapons and human beings.

When used on machines, energy weapons are called High-Powered Microwaves: when pointed at human beings they are called Active Denial Weapons. Because energy-beam weapons can be tuned, goes the argument, they will not be used to cook humans but to 'actively deny' them the ground they occupy. Demonstrated on military volunteers (in itself an intriguingly nauseating fact), ADWs drive people away by heating the water in their skin to unpleasant (read unbearable) levels. The sweeping beam can be used to push crowds back, herd them towards or away from areas. In love both with its euphemisms and with its linguistic realpolitik, the military unofficially calls Active Denial technology 'Dial-a-hurt' weapons because, as with microwave ovens, the power can be turned up until the dish is steamed. In 2001 U.S. Senator Pete Domenici (R., New Mexico) read his reassurances about ADWs into the *Congressional Record*, arguing that 'the energy cannot be "tuned up" to a level that would immediately cause permanent damage to human subjects. The technology does not cause injury due to the low energy levels used. ADT [Active Denial Technology] does cause heat-induced pain that is nearly identical to briefly touching a lightbulb' (Domenici). Voluntarily touching a lightbulb is one thing – spraying yourself with a welding arc is another. In Domenici's soothing words are hints that plenty of damage is there to be done. People caught in the beam, people who have fainted, cannot move, or are wounded, will suffer more than if they had touched a lightbulb – they will experience damage beyond the 'immediate.' It is entirely unclear how such beam weapons act on the brain or the central nervous system, during either short or long periods of exposure. New plans for ADWs include beams that would 'disrupt a person's short term memory and cause him to lose control of involuntary body functions' (Vizard *Scientific American*). The Air Force Research Laboratory's HPM information sheet notes that 'whereas a typical microwave oven generates less than 1,500 watts of power, the Division is working with equipment that can generate millions of watts of power' (USAF, Air Force 'High Power'). To return to Senator Domenici's analogy for a moment: given that the average lightbulb is sixty watts, a weapon that can focus millions of watts of power in one place is a very large and dangerous lightbulb. If it is impossible to stay away from the image of the microwave oven, that's because HPMs and ADWs are portable flesh-fryers useful against computers and humans indiscriminately. There's nothing to stop a UCAV from firing hundreds of microwave pulses into crowds of people (except perhaps other people who decide that doing so isn't a good idea). HPM weapons are now in use. Just before the second Gulf War started, a reporter who asked an Air Force spokesperson about plans for HPMs in Iraq got the answer: 'I can't say much about [readying HPM weapons for use in Iraq] ... We've worked in high-power microwave areas and continue [to improve them] in terms of power output' (Fulghum and Wall 'Deployment' 37). The lack of Iraqi air power, the rapid and thorough collapse of the electrical grid, and the absence of surface-to-air missiles fired at American aircraft speak silently about HPM weapons. The Air Force narrative lines up perfectly with the weapon's name: it is an 'active denial' that communicates flawlessly. Not saying much tells us a lot. As with everything about UAVs, the projects are black, in the dark, staring ruthlessly at humans who are fixed in the electro-optical headlights of real-time streaming video, pinned to the earth by synthetic aperture radar, persistently and actively denied ground, air, and life by beams of energy they can neither see nor avoid.

The light, high-tech forces desired by RMA planners are a perpetual nightmare for a military fearful of obsolescence. Acronyms for robotic ground (UGV), water (UUV), and air (UAV) vehicles have in common the letter U for 'unmanned' or 'uninhabited.' Institutional resistance to shift from individually guided heavy weapons to interoperable uninhabited light vehicles is thick: just as cavalries once refused to exchange their horses for ungainly tracked steel boxes, and officers reluctantly let go of sword hilts in favour of automatic-weapon gunstocks, now the whole military must confront the surrender of Warsaw Pact war-fighting. For many soldiers, continuing encroachment by uninhabited craft represents the loss of work, the loss of agency, the loss of glamour associated with combat. Pilots in Vietnam saw in the Firebee drone a machine opponent that stole their flying time and its associated combat pay. Lloyd Ryan, whose company built and operated the Firebee, and under the aegis of Northrop Grumman now builds the Global Hawk, notes: 'It was a problem ... to explain that a low-to-medium altitude unmanned drone system was not in competition with the big, fast, high-altitude piloted airplane – that it really constituted one of several capabilities the Air Force should have and continue to develop' (Wagner 16). The song the uninhabited industry must sing has a 'both-and, not either-or' refrain: uninhabited vehicles are always talked about as force multipliers supplementary to humans, not as forces in themselves.

The Firebee was referred to as an RPV, or Remotely Piloted Vehicle, not an uninhabited aircraft; preserving the word 'pilot' made it clear that humans controlled the command loop. Ryan engineers spent years in Vietnam (some even drew combat pay) and flew the Firebee RPVs, finally handing control over to the men they called the 'blue-suiters' (Air Force pilots and technicians). Only when Air Force pilots began to be shot down in record numbers over North Vietnam, only when Ryan Aerospace was able to get combat pay for remote pilots, were the drones grudgingly supported. The struggle for legitimacy isn't over: the once celebrated air ace is now a mouse jockey, pointing and clicking on a screen dozens or hundreds of kilometres away from combat. Despite UAV performances in Afghanistan and with the Israeli Air Force, traditionalists remain irritable: '"Could we build a machine that does an appendectomy? Absolutely. Could it do it very well? Absolutely. Would it be expensive? Absolutely. So why bother" replacing the doctor, [Roche] said. The same reasoning may hold when considering replacing the pilot' (Wall 26). Two kinds of discourse are at work: the first belongs to the world of cost-benefit accounting, a discourse locked into place by Robert McNamara's war on Vietnam; the second comes from the world of professionalism. The warrior is actually a professional healer, a surgeon for the body politic who cuts away a ruptured organ that endangers the whole. Everything is possible – 'Absolutely' – 'Absolutely' – and everything can be accomplished by human reason and science, so keep the real brains in charge. Trust people, consider machines only as tools (the surgeon needs a scalpel in order to perform the appendectomy), and, setting aside the concern for human warrior-healers, remember that broken humans are still cheaper to replace than wrecked machines. Yet, below the official radar, one can still hear the hissing interservice static suggesting that humans will lose their place on the battlefield.[32]

Even pilots who accepted UAVs couldn't resist competing with them. In 1971, seeing a chance to show the Air Force how jets and UAVs could co-exist, one Ryan official challenged John Smith, then commander of the Navy Fighter Weapons ('Top Gun') School to fly against the Ryan drone and kill it. While the flight began with a cheery chivalry ('"Tally ho, off the left wing" called Smith as he tried to line up for the kill, but the drone was able to pull such a high-G turn that the F-4 could not follow the maneuver ... "It's turning like a mother"'), it ended somewhat glumly for Smith as the drone survived unscathed. What went unspoken was that humans simply could not and cannot withstand the physics of the manoeuvres an uninhabited craft can perform. As one of the Ryan engineers reports, 'Smith was learning the hard way that we could rack the Firebee into a hundred degree bank and make a 180 degree turn reversal in only 12 seconds,' which we can translate simply: Smith began with the drone in his sights but ended up with it on his tail (Wagner 188). The limits of the flesh, the pressures of blood flow, the softness of eyeballs, the strain on the heart, make it impossible for humans to operate at greater than six or seven gravities for any sustained time (one gravity ['G' or 'gee'] is Earth normal, as if one were on the Earth's surface). Future UAVs and UCAVs will be built to withstand a consistent twenty to thirty gravities, three or four times the Firebee's (and a human's) endurance. Smith's good-natured loss doesn't erase human unease about the accomplishment: by virtue of his position at Top Gun, Smith was one of the best pilots in the United States Air Force, flying against a craft that is now some thirty years old. And he lost.

By the summer of 2002, humans were not just unable to keep up with UAVs – they actually damaged UAV performance: 'At the speed at which things are done and the complexity of the signals that you're trying to work on, increasingly there is less value added by a man. A man just slows things down. Distributed systems are a lot more complex, but inherently you're much better off with unmanned' (Fulghum and Wall 'USAF Tags' 26–7). That's Mark Ronald, president of BAE Systems North America, director of a company entirely reliant on defence contracts for its uninhabited war craft (missiles, UAVs). His vision is that of the RMA: distributed systems, where no node is more senior to another, where the system is goal-oriented, requires unhesitating 'throughput.' Humans may direct information based on ego and unpredictable chemical sensations, but systems of systems don't dally with cogitation, fears, and second thoughts. There is no importance attached to the node that provides the key information or solution: all nodes are equally unimportant, or important. Ronald's rhetoric neatly embosses the human's ambiguous position of being 'increasingly less'; humans create an increasing absence.[33] Netcentric warfare sees all information sources as equal: if a piloted craft, high and low ranks, and an autonomous machine operate on the same level, then perhaps the demolition of that bane of military communication – the information stovepipe – has finally occurred. In the Warsaw Pact environment, a private in a mechanized unit would pass information up to a commanding officer, who would move it further up the chain to central command, where it might or might not be redirected down another stovepipe to an Air Force commander, who could then connect with a pilot. In the *New World Vista* interconnection fantasy, all stovepipes have been shattered and information flows to the person (or machine) that requires it. Military-communication models some hundreds of years old will, apparently, have been razed.

Limitations on UAVs created by low bandwidth have gradually been pushed back. Keeping in mind Intel chairman Gordon Moore's 'law' that information density on a computer chip doubles every eighteen months, the defence establishment waited for faster computers (that is, machines with high clock speeds, able to handle data at very high rates) and large-bandwidth wireless modems.[34] High transmission and reception rates, mass-data storage, and swift retrieval capability make the UAV a self-evident device. When uninhabited craft were Remotely Piloted Vehicles, the guidance and operation systems had to be stored in the brains of nearby human operators; information could not be processed quickly enough by an RPV for it to be cut loose from its human moorings. Now that a high-end desktop computer can store half a terabyte of information (a terabyte is a thousand gigabytes – most computers now arrive with 120-gigabyte hard drives), UAV brains are nearing the size and clock speeds at which they can be self-sufficient. UAV independence is just fine, according to the *2002 Unmanned Aerial Vehicles Roadmap* (commissioned by the Office of the Secretary of Defense), which says bluntly: 'The more intelligence we can "pack" into the UAV, the more complicated task we can assign it to, and the less oversight required by human operators. We must continue efforts to increase intelligence of these vehicles' (131). The human spinal cord will shortly be severed and the machine brain will assume command. Humans will set up, maintain, and give general guidance to UAVs, but the rest will happen onboard, in the air. The distinction between intelligence and computing power doesn't seem to be an issue: Does high-speed computing, data transfer, and complex multitasking make a machine intelligent?

The bottleneck is now on the ground. Humans can't easily handle the amount of information UAVs can send. A human operator may be overwhelmed by simultaneous visual data along with demands from multiple sectors asking for target information, changes in course, and for weapon decisions: Is it the right time to fire? What weapons should be used? The calm of pre-planning sloughs off as the battle heats up – it isn't the fog of war so much as its vapour, the sweat and vertigo caused by unmade decisions, bad decisions, or too many decisions. In order to bank the flood of data drowning the UAV controller (a reduction critical for the operator of a 'weaponized' UAV or UCAV), the Navy Aerospace Medical Research Laboratory (NAMRL) has developed the Tactile Situational Awareness System (TSAS). At the moment the system is only a vest (a full-body TSAS seems likely to be forthcoming), which uses touch to advise the wearer of the craft's status. The Navy's website (with its utter lack of discussion about the program – a deep presence of absence) shows a rather ungainly cartoon human wearing what looks like a quilted vest (fig. 17). Each of the 'buttons' on the quilt signifies what the Navy calls a 'tactor' which, driven by air, taps the wearer, depending on the amount of yaw, pitch, or roll the craft experiences. Because the system works on the tactile level and can be learned reflexively, it will free the operator to do high-level cognitive work, processing data and making decisions based on it (the TSAS has already been successfully tested).

The TSAS unloads the brain but extends the human by literally putting more pressure on the body. The body is slaved to the machine: what ills the machine feels, the human must interpret and adjust for. The machine doesn't help the human, except by removing the human from the battlefield, and in exchange the human becomes a new kind of drone in the network. The drive towards telepresence, for the user to be at the scene yet not in danger simultaneously, picks up speed. Early drones were programmed before launch, then flew, ditched, stayed on course, returned, or didn't. Late in the Vietnam War, drones were remotely guided by operators flying nearby in larger aircraft, as much out of harm's way as possible. The TSAS reinserts the opera-

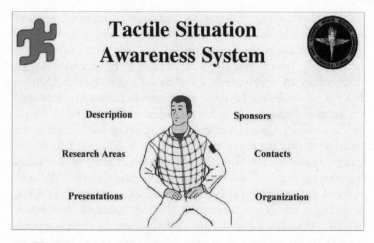

17   **Fly Me**: the tactor suit keeps the wearer's mind free as the body becomes a replica of the aircraft. Note the logo in the upper left – the human as puzzle piece that must fit the interoperable environment.

tor's body, reconfigured as a machine, into battlespace. The cheery looking Ken-doll cartoon TSAS wearer, bemusedly surrounded by vacant or dead-end hyperlinks, obscures the worrisome issues of crossed boundaries, the convergence of the tactile and cognitive, the need for data processing at a speed that other sensory systems (optical and aural) can no longer sustain. Tellingly, *New World Vistas* recognized that the 'thrust of entertainment technology is to convey a sense of "being there" to an audience or to a group of participants,' concluding 'We urge the Air Force to establish continuing contact as closely as possible with entertainment organizations' (USAF Scientific Advisory). Simulation is inseparable from high-technology industrial warfare, and the military's commitment to its more than dozen so-called 'battlelabs' that can run extensive multiple-user war-fighting experiments shows that it accepts simulation as part of war. But the Program Executive Office for Simulation, Training, and Instrumentation sees its mission in more transcendent terms, as its motto, the now infamous 'All but war is simulation,' indicates (fig. 18).

There are a number of claims made by the PEO STRI motto: that

18   **Making It Real**: The official flash for PEO STRI reflects the clean world of the simulated war-fighter. The red bull's eye is converted to a safe blue, the two halves of the world divided by battlement crenellations (the soldier is now safely inside simulation's fortress). The lightning bolt (signifying cybernetic operations) connects the two worlds. Around and inside are the safe green laurels – the anonymous silhouetted soldier will be preserved by the enfolding leaves of simulation.

war is the first and most important reality (the rest is unreal), that even simulated war isn't real war (although remote UAV operators wouldn't be able to tell if they were watching a film of combat, or actual combat, on their monitors), and that peace, that which is not war, is fake. Even in a time of what are called Low Intensity Conflicts or Operations Other Than War, the PEO STRI phrase makes it clear that an era of total war has arrived: only war makes reality. Still, the concept of far-flung netcentric war was met during the 2002 war on Afghanistan with characteristic irascibility by Air Force General John Jumper, who barked at a reporter: 'I'm not going to undertake this amorphous network centric thing without a verb in it ... I'm going to start with the thing that kills the target and invent the network that locates it, identifies it and instantly get the information to the person or the warhead that is going to blow it up' (Fulghum 'Network' 33). Of the Air Force generals, Jumper has been one of the strongest proponents of the RMA, even to allow that UAVs will fly, point, and kill. Information is important only as long as it leads to the 'person or the warhead' (suggesting an interchangeability between the two) and then to a kill (Jumper's verb of choice). Jumper may embrace interoperability, but you can see his lips peeling back from his teeth as he does so.

The brilliant and erratic, troubled, tragic, science fiction author Philip K. Dick saw more clearly into the oddness of human-machine interactions than many other post-war writers. Through Cold War frost, Dick envisioned a world mired in what war theorist Paul Virilio would call total peace, a nightmarish, totalitarian condition in which humans are entirely, as Thoreau would say, 'tools of their tools.' Dick's whimsical and unrelenting story 'Autofac' (written in 1954 and published in 1955) shows automated factories busy plundering the Earth for raw materials in a post–World War Three environment: the hapless survivors find that they cannot shut the factories down. An attempt to disrupt the machines causes a resource war between competing factories, and soon humans are forced to hide in the ruins of ruins as Von Neumann replicators (John [Johann] von Neumann proposed intelligent machines whose first task on being activated would be to replicate themselves) slug it out with autonomous vehicles on land and in the air. When one set of machines serves another, it's hard not to recall Dick's prophesies.

In the second Gulf War, UAVs were used, as they had been in Vietnam, to expose missile sites. In acting as lures for missile-targeting stations, drones and UAVs reveal launchers that are then attacked by air-to-surface missiles: 'One Predator plunged into the Tigris and the second into a lake. The UAVs were referred to as "chum" because they served as bait for the anti-aircraft defenses' (Fulghum 'New Bag' 22). Two Predators (there are said to be some seventy currently in operation) were sacrificed for the good of the mission. 'Chum,' fish refuse that bloodies the water and attracts sharks, is provocative. Just before the end of Dick's autofac war, humans watch a showdown between an intelligent ore cart and a hovering airborne hunter-killer called a 'hawk': 'In the sky, the hawk swept in a straight line until it hung directly over the cart. Then, without a sound or warning, it came down in a straight dive. Hands to her face, Judith shrieked, "I can't watch! It's awful! Like wild animals!"' (Dick 16). The hawk's strategy pays off, for, while it is immolated, it manages to wedge open the enemy autofac gates and set off a bomb inside. Rather than anthropomorphize them, the humans understand the machines as beasts; Dick brings them to life with a few adjectives (he describes the damaged ore cart as moving 'painfully, with infinite caution'). Humans have become agentless spectators on a dying planet.

Dick's story warns not only of war machines that burrow into the Earth's heart but also of the ecological cost of unbridled consumerism. With the Korean War, famous for its human-wave attacks that were the antithesis of the autonomous interoperable battlefield, at a standstill, Dick's story must have looked like utter nightmarish babbling. But uninhabited vehicles have since greatly increased their liveliness, and the *2002 UAV Roadmap* lays out the evolutionary ladder for future UAVs in which remote piloting (teleoperation) has long been surpassed. The curve soars from UAVs that will be able to reroute themselves when confronted with battlespace changes (level four), coordinate with other uninhabited craft and replan tactics as a group whether on the ground, in the ocean, or in the air (level six), and pursue various 'distributed' goals independently but with the same end in view (level eight), the pinnacle of operation will be: 'fully autonomous swarms' of rolling, flying, and swimming UVs, planning and executing their own strategic, not merely tactical, goals. Currently, the

Global Hawk, smartest of the fully operative machines, ranks a lowly 2.5 on the scale, sort of like a flying 1931 Maytag wringer-washer.

Global Hawk is long-loitering (can stay over an area for more than a day before returning to base) and is told only 'where [the operators] would like to go and what they are interested in looking at; the rest of the equation is worked out by Global Hawk on its own' (McDaid and Oliver 119). Even without the TSAS telepresence suit, the UAV has already carried off its users who submit to its route and flight plan. Global Hawk's autonomy will increase as computers become faster; hundreds of terabytes of high-speed onboard memory will allow the craft to alter its plans without hesitation. Unwittingly following Dick's predictions about the future battlefield, the *2002 UAV Roadmap* calls for uninhabited vehicles of different families to speak to and take orders from each other: 'Future UUVs [Uninhabited Underwater Vehicles] may themselves deploy UAVs to extend their capabilities and improve overall system performance. Small UAVs that become unattended ground sensors will blur the distinction between the classes of USs [Uninhabited Systems]' (64).[35] In order to preserve themselves, UUVs will need to see over the horizon and will launch micro-UAVs like the Hornet that then report to the UUVs and the network; were the Hornet to perch and stare, it would become a ground station. Uninhabited vehicle boundaries will be further smudged by so-called 'parasite' UAVs launched from larger uninhabited vehicles; such metadrones have already been used extensively to act as advance guard for the advance guard. Micro-UAVs can be gun-launched, dropped, and spun off host UAVs (in the early 1990s 'mesicopters,' tiny flying disks with rotors the size of a penny, were designed to form disposable mesicopter swarms). The RMA's interest in battlefield swarming, aroused by the Rangers' 1993 experience in Somalia at the Battle of the Black Sea, has led to the desire for massed UAVs that can launch secondary UAV swarms. One variant of the secondary UAV is the Loitering Electronic Warfare Killer (LEWK), a one-way recoverable UAV designed to stay on station until it sees its target. In Afghanistan, eager to watch the many cave openings that might house enemies, the Air Force asked for Predators from which it could drop LEWKs, which would then loiter and attack at any sign of movement; the Predators would have already moved on.

As UAVs become more prevalent, smarter, quicker, it seems likely that ground controllers will grow accustomed to them. But sooner or later human ability is going to fail, and an unnoticed UAV is going to commit what is blithely known as a 'blue-on-blue' incident (that is, friendly fire).[36] A UAV is going to kill one or more friendly soldiers (as opposed to civilians – another discussion entirely), and even if the story doesn't reach the public, panic will set in. Humans will be pushed back into the loop, even if only temporarily. But the drive towards autonomous strategy-creating machines won't stop. The focus will continue to be on UAV health. UAV classes (in the economic sense) have already formed: a number of Ryan Firebee drones can be sacrificed to save a Predator, the Predator can be dumped into the Tigris if it paves the way for a Global Hawk or the most expensive of the Air Force's jet panoply, the $35-million Joint Strike Fighter complete with pilot crammed full of another dozens of millions of training dollars. Even so, UAVs won't be sacrificed wantonly: the vehicles will be hardened against attacks of various kinds: 'Vulnerable sections will be wrapped in Kevlar, man-rated engines will be installed, avionics will become triple-redundant, and critical components will be separated to avoid simultaneous failures' (Fulghum 'Predator's Progress' 49). In order to protect the investment, the UAV will begin to receive the kind of attention once reserved for humans (Kevlar clothing) and their craft: power trains that are more robust, back-up pieces in case of lemons. High-Powered Microwave weapons are intended for UAVs in part because the fear is that a jet firing thousands of microwave bursts might knock itself out of the sky (the shielding around an HPM must be immaculate). Similarly, Electromagnetic Pulse (EMP) bombs, designed to produce blasts of dirty radiation that fill the radio spectrum with information junk, may wipe out aircraft avionics. One dream of survivability is to provide each UAV with nothing less than a map of the world at one-metre resolution. Such a map, the authors of *New World Vistas* conclude, 'will require 10–20 terabytes with suitable compression. After the creation of the initial map, only updates need be supplied routinely,' so that all UAVs see what each sees. The report continues: 'We refer to the high resolution onboard digital map as the "onboard world"' (*New World Vistas*).[37] Such an instrument is in fact a new planet over which slides interlocking systems of systems, where unattended vehicles roam the land, fly high, perch and stare, and troll the littorals and the deep ocean, 'messaging' each other in Internet chatrooms (the Predator was controlled through chatrooms in the second Gulf War) to which humans may, or may not, have keys.

The defence establishment has created a micro fog of war for UAVs, valued for their blackness, their unknown qualities. Some details about UAVs have been released to let the enemy (variously constructed) know that they can be seen – panopticism doesn't work if the inmates aren't perpetually reflexive. Near-future plans for UAVs include their use for patrolling the Rio Grande as well as American coastal waterways (the Coast Guard plans to fly a Broad Area Maritime Surveillance [BAMS] system of UAVs in keeping with the Orwellian Homeland Security program). In 2003 UAVs were granted license by the Federal Aviation Authority (FAA) to fly in American civilian airspace: that means uninhabited craft can operate outside military bases and proving grounds, near civilian air traffic, over cities. In late 2003 the 'Access 5' program united leading UAV makers, NASA, the Department of Defense, and the FAA to open national air space to UAV operation within five years. What craft will fly, how resilient they will be, what distances they will cover, we don't know. When it comes to the details, there's a lot the public, or military researchers, don't know about UAVs. Like the craft, we're in the dark. Of two well-known books about UAVs, one was held back from publication for ten years (the Pentagon intended to censor it so heavily that nothing much would have been left to print [Wagner 'foreword']); the other contains a disingenuous note in the front matter by the then chief of the Defense Airborne Reconnaissance Office (the Pentagon) about the authors' 'speculation' over 'unsubstantiated activities at clandestine military facilities' (McDaid and Oliver 9). Apparently, the motto of most UAV researchers should be: 'All but war is speculation.'

What is not speculation is the Pentagon's July 2003 confirmation that it would proceed with a hypersonic UCAV (most are subsonic) that 'could strike targets anywhere on earth from the continental United States (US) in less than two hours' (Shepard's 'Pentagon Plans'). These machines are not missiles, it's worth remembering, but high-flying returnable aircraft that will have to be capable of maintaining speeds in excess of Mach 5–6 (roughly 5,600–6,700 kph) to accomplish such stunts. The drive for speed, for strike at distant targets, for global reach and grasp, is in keeping with the rest of the RMA's manifesto. The wars on Afghanistan and Iraq confirmed for UAV and RMA believers that uninhabited vehicles are critical to future war-fighting (the group is always fixed on the same things – information warfare, systems of systems, domination of the battle-space through netcentricism, screamingly high and expensive technology, electronic attack, jamming and spoofing). New confidence in the RMA's mission can be seen in the way UAVs were handled during the March 2003 attack on Baghdad when 'two ground-launched and three air-launched Northrop Grumman BQM-34-53 Firebee drones were used to set up chaff corridors to shield manned aircraft and cruise missiles' (Fulghum 'Targets' 54). We're accustomed to seeing the Firebee drones slung into battlespace like hot rocks, but these drones were used in an autofac fashion: they littered the air with chaff (strands of metal that flutters as it falls, preventing ground-missile sites from locking on air targets), forming a sacrificial drone-aircraft clear zone for cruise missiles. Commenting on the use of drones as battlespace pawns, a diplomatic Northrop Grumman official who remembered Firebee history noted cautiously that 'it was not that anyone forgot what had been done during Desert Storm and Vietnam, it was just that there were new people [planning the conflict] who never knew you could see targets from drones that way' (Fulghum 'Targets' 54). In other words, it wasn't forgetfulness but just plain old ignorance at work (it's hard to decide which is more worrisome). What the speaker, Douglas Fronius, director of target programs at Northrop Grumman, avoids saying is that the new generation of weapons' users didn't have the imagination to use drones as aerial shields for the next assault wave, to use drones for drones. Institutional forgetfulness, particularly when the institution concerned is the military, is unnerving; but lessons *were* learned, and a new generation of UAV users now know how to begin thinking about systems of systems.

Long-striking, high-flying UCAVs clearly top the UAV shopping list in *New World Vistas* and the *2002 UAV Roadmap*. If UAVs don't currently have the speed desired by the Pentagon, they have overcome both height and distance. Ironically it is AeroVironment, the same company that makes some of the smallest, shortest-range craft (Black Widow, Wasp, Hornet), that developed the biggest, highest-flying UAVs, Pathfinder and Helios. AeroVironment's Helios (fig. 19), a joint project with DARPA and NASA, is a solar-powered fully autonomous flying wing (the top surface is completely covered with solar cells). Before it crashed in June 2003, Helios had flown to almost 300,000 metres, and loitered in the air for seven hours. It would next have flown unattended for forty hours. Helios is part of a program that AeroVironment calls 'Skytower Global,' a nineteen-kilometre-high flying tower providing phone and Internet service, as a satellite does. On take-off and in flight, the craft is astonishing. Because of its fourteen lazily turning propellers (slow enough that individual turning blades are visible), and seventy-four-metre wing (by comparison a Boeing 747 jet has a sixty-three-metre wingspan), the Helios has a leisurely, dreamy quality about it. Videos show it rising off its tropical testing field, meeting thermals with the poetic flux of a manta ray coasting through air. The wing undulates gently, adjusting itself as it murmurs steadily onward. Rather than use tropical snapshots, AeroVironment chose to publicize Helios with an image of the craft emitting a cone of light while flying over New York City (the World Trade Center, or its lack, isn't visible, but the message is clear). The ad copy emphasizes the craft's communications ability, but the picture signifies a mechanical angel spreading its umbrella of protective information technology. Somewhere overhead is a slow, long-loitering, stratospheric traveller, an autonomous flyer that can dwell over its subject for almost two days. The staggering images of the Helios in flight are tempered by the vision of it casting its gaze over New York, just as the Predator oversees the whole globe.

A solar-powered flying wing doesn't make possible the kind of speed that a deep-strike UCAV needs; for that reason the *2002 UAV Roadmap* has restarted the push for nuclear-fuelled craft. Nuclear-powered UAVs would have the endurance and lifting power required for globe trotting, but, as the report reluctantly concludes, 'the principal limitation on development of this technology will likely be the policy implications associated with operating a nuclear reactor in

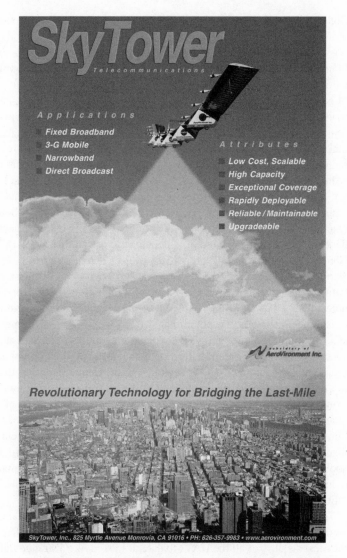

19 **Let Information Techology Be Your Umbrella**: AeroVironment's Helios goes on forty-hour unwatched guard nineteen kilometres over New York City.

flight over foreign countries' (Office *Roadmap* 118). Or *any* country.[38] Neither the *Roadmap* nor *New World Vistas* intend to put a flying nuclear pile overhead. The technology under discussion as of February 2003 is a 'quantum nucleonic reactor' that operates by bombarding hafnium-178 with X-rays, which produces energy in the form of gamma rays (heat), then used to propel the craft. The nuclear UAV would be able to dwell autonomously not for days, but months. In a standard reactor, atomic fission in the uranium core creates heat that makes water into steam which drives turbines (the water in the core is radioactively 'hot,' and the containment area rather rapidly becomes radioactive as well). In a quantum nucleonic reactor, as soon as the X-ray bombardment stops, so does the reaction and attendant power emission. There can be no danger of meltdown or runaway chain reaction. That's the RMA position. But this is another black project: dark matter occludes quantum nucleonic reactors and hafnium-187. Challenged on hafnium's properties, project scientist Christopher Hamilton at Ohio's Wright Patterson Air Force Base was dismissive: hafnium is not particularly radioactive, he argued. But other scientists noted that it has the same half-life as the staggeringly dangerous caesium-137.[39] Hamilton noted that, when the X-ray bombardment stops, 'gamma ray production is reduced dramatically,' perhaps unaware his listeners wanted to hear that gamma-ray production would cease entirely, not merely be lessened. Pressed, Hamilton agreed that the reactor was 'probably something you would want to stay away from but it's not going to kill you' (Graham-Rowe). The image of a quantum nucleonic reactor-powered Global Hawk shutting down in mid-flight and sliding toward earth, spilling its dirty carcinogenic load over miles of land or water as it crashes and breaks up, is hard to abolish. We may also recall that employees at Northrop Grumman didn't know about drone use from Vietnam or the first Gulf War. Institutional blackouts and UAVs seem to attract one another.

The most energetically forgotten UAV story is the DarkStar's. Built in a Lockheed Boeing partnership, the DarkStar was perhaps the most stealthy of all the UAVs to date. It looked like a strange flattened disk attached to twenty-one metres of wing. Already blindingly overbudget, the DarkStar was on its final, fatal flight when it tipped, caught a wing on the tarmac, crashed, and burned. The pieces were examined but 'the remains were then placed in a steel container and its doors were sealed and welded shut. The crash and subsequent fire had apparently made some exotic structure, probably the anti-radar coating, toxic. The DarkStar coffin was then buried at an undisclosed site within the Edwards Air Force Base boundary' (McDaid and Oliver 121, 129). Mentioning DarkStar in the UAV world is like uttering the word *Macbeth* backstage at a theatre. Banquo is persistent, however, and, late in the second Gulf War, an unusual UAV was seen flying over Iraq. Questioned closely, the Air Force finally admitted that there *was* something there, some new kind of UAV, something with 'the same concept as DarkStar, it's stealthy, and it uses the same apertures and data links ... The numbers are limited. There are a couple of airframes, a ground station and spare parts' (Fulghum 'Stealth' 20). These few sentences were all even the ingenious *Aviation Week* was able to elicit or, more likely, print. Forgotten drones, acceptable radioactive hafnium you 'would want to stay away from,' a buried coffin with a vampiric UAV inside: Philip Dick already knew what was coming.

At the very end of his story about the automated factories that erase human life, the future, and Earth's ecology, Dick recounted the death of the warring autofacs. Near one of the factory doors the characters find a tube periodically firing out a canister that cracks open on landing. Inside the canister is a self-contained unit with 'microscopic machinery, smaller than ants, smaller than pins, working energetically, purposefully – constructing something.' Since they are Von Neumann machines, they are building new autofacs: 'The bits were in motion,' Dick says. The horror of the self-replicating system of systems, long-dwelling machinery, is allayed by Dick's sarcastic laughter at his dull-witted schoolboy men who are thrilled: 'That would be neat – autofac networks through the whole universe' (20). Through the ether over the planet's dark and light sides, moving through the troposphere, others muttering about near the ground or in upper airless space, swinging in wide slow circles, uninhabited aircraft loiter and dwell, chat on their Internet, passing over the landscape all-seeing, seeing all, and dead slow.

Chapter Six

Wastage:
War after War

The war has been like a nail in my head, like a corpse in my house.

– Larry Heinemann, interview

What if an angel of truth
had come to us and said
'Enjoy your last whole day
Tomorrow you'll be dead.'

   And pointing at me:

'Except for you.
You enjoy
your last day whole.'

– Joe Haldeman, '13 September 1988'

Back on the ground; maybe in it – dirt in the mouth. Travels low in rotorcraft or lofty uninhabited airspace are over, no more flying. The state relinquishes the body, wars' cicatrices inscribed on the flesh by visible and transparent forces. There may have been some losses: it could be hard to walk, to grip, to see, to think. Could be impossible. Limbs have vanished, feelings have slipped, bits of the past are now missing, sanity has loosened a little: panic holds a place where comfort was (being inside, that's hard – the dark, loud sudden noises, crowded public places, eating, talking, sex – they're tough now and once they weren't). The whole, a picture of how to live and why, has been replaced by shatterings. Uncertainty is troubled by spikes of panic, waking and sleeping nightmares. The thousands we know (58,000 in one day on the Somme, 60,000 at Stalingrad, not to mention the 40,000 in the Blitz, 135,000 at Dresden and other charnel cities), the thousands, hundreds of thousands, tens of millions, forgotten in industrial wars we may not recall, challenge us to count and make it mean something (no false cries of 'never again' when we're still prepared to say 'unless tomorrow'). Apart from the ones who were there, who can tell the difference between sixty and eighty thousand? Which ones are *they*? Where the bodies can't be recovered, the mass graves make for lush harvests. The broken can be carried home and struggle to stay upright on legs become metal pins, pincers for hands, plastic doll limbs for cosmetic occasions. Ghost armies wander through civilian life, battalions of people who look backward: some will never turn around again. All the while the postwar world counts: the suicides, the amputees, the panic-stricken, the payouts made to walking scar tissue. It's all part of the same measurement scheme that has calculated the armoured suit's production cost, the tank's glacis angle, the rotorcraft's agility, the UAV's persistence. Anxiety and pain make the body hot with wordlessness, a perpetual seizure, the force of earth pressing on the chest wall, squeezing out the breath, while the living world tries to step away from the abyss. Forgetting is important now. Forgetting the ones who are too damaged to do anything but cling to the edges of life. Let them fall: they're waste matter.

During the First World War, the British High Command, inexperienced with slaughter of such magnitude, came to call the daily trench casualties 'wastage' (Shepard 45). 'Wastage' – linguistic dirt thrown on the corpses of those killed in planned battles and steady theatre-wide losses (various causes of death: artillery, gas, gunfire, random accidents, sickness, loss of heart); 'wastage' converted intolerable numbers into meaningless ones (meaningless to war accountants that is, not the dead themselves, their relatives, or fellow soldiers). Lives were spent in seemingly endless battles of attrition that locked up war zones for months at a time, and in set-piece planned disasters that took all soldiers regardless of their nationality, like Loos (80,000 killed or wounded between the two sides), the Somme (over a million), Passchendaele (over half a million). Only 'banking language,' as Ben Shepard calls it, could properly assess the costs of industrial war: to determine a win it was now necessary to establish a balance sheet and charge ground gained against the cost in bodies. The same language would be used even more aggressively by post-World War Two managers like Robert McNamara, focused on his daily Vietnam body count; if you live by the tenets of wastage, the dead will be tallied as proof of victory, or at least progress.[40]

Industrial war and its products focused the military on men it couldn't afford to lose: those no longer able to bear the trenches. Body-units in the war factory had to be present and reliable in order to function as High Command wished. Soldiers who froze or became so panic-stricken they could no longer operate were not, as was first thought, of poor genetic stock or class. Background, breeding, training, experience – nothing seemed to explain why some could continue and others could not. Perhaps it had to do with high-explosive shell bombardments, perhaps a force that affected the inside of the human body, something that early psychologist Charles Myers called 'an invisibly fine molecular commotion in the brain,' a shell that produced a shock to the nervous system (Holden 17). Inside the body or, worse, the secret turnings of the brain's maze, something was going wrong. Because weapons could persist as they never had before, because apparently endless supplies of shells, machine-guns, accurate rifles, grenades, and gas kept the front operational around the clock and the calendar, soldiers could no longer find relief from their work. Long marches punctuated by day-long or even three-day battles (Waterloo, 1815) were transformed into protracted slaughters measured in hundreds of days. An unexpected result of extended killing was men who stuttered or couldn't speak at all, who couldn't walk, see, or taste, yet whose flesh was unbruised. Helpless before these blank bodies, physical medicine grudgingly surrendered its 'neurological' cases to psychiatrists who, it was hoped, could perform the work expected of all wartime medicine: healing the wounded and ill so they could return to the factory that is war.

Between 1914 and the mid-1960s, the state increasingly looked to psychiatry to save it from the enormous post-war costs of caring for mentally stricken veterans. British and Americans were still paying out First World War pensions even as the Second World War began (40,000 cases had still to be acquitted in Britain). A monetary accounting of the expenses future sick veterans would incur drove the state, through military psychiatry, to screen recruits; should the Allies win the war, it would be better if few or no pensions were paid to psychiatric casualties. Screening looked for men who seemed stable, perhaps a little dull, even lacking in intelligence, but also free of the imagination that psychiatrists felt created most war neuroses. Perhaps men raised in the country would be better suited to combat stress than urbanites, who were, on the whole, perceived to be neurotic. The Second World War psychiatric weeding process resulted in a military disaster when half a million men, enough to staff fifty divisions, were pulled aside (they were later called 'the lost divisions' [Shepard 326, Ginzberg]). In 1966 an attempt was made to recapture some of these numbers when, in accordance with Lyndon Johnson's Great Society Program, the army dropped its entrance standards. Then Secretary of Defense Robert McNamara oversaw 'Project 100,000,' designed to bring 100,000 men (mostly poor or of colour, or both) into the military where they would receive skills training. Project 100,000 (also known as the 'New Standards Men' because the standards had been lowered) ultimately netted over 300,000 people, many of whom were fodder for Vietnam. But as the war persisted and national disenchantment with it grew, psychiatry began to break with the state: 'Prior to the 1970s, [the psychiatric] agenda involved colluding with the military in order to "salvage" and return agitated soldiers to combat ... With Vietnam, however, one sees a change in emphasis' (Dean 44). Psychiatrists like Robert Jay Lifton and Chaim Shatan, themselves openly anti-war, criticized the military mission and encouraged their patients to do the same.

Joseph Heller's famous 'Catch-22' elegantly concentrates the position of Second World War military medicine: 'Orr would be crazy to fly more missions and sane if he didn't, but if he was sane he had to fly them. If he flew them he was crazy and didn't have to; but if he didn't want to he was sane and had to' (Heller 56–7). There is, as Heller says, a terrible 'spinning reasonableness' to this calculus, of which none remains in Vietnam veteran Gustav Hasford's *The Short-Timers* (1979) (filmed by Stanley Kubrick as *Full Metal Jacket*

[1987]) and its sequel, *The Phantom Blooper* (1990), where 'a Navy psychiatrist is to psychiatry what military music is to music. No fucking pogue lifer questions Command.[41] Even the chaplains are on the team. The job of the military psychiatrist in time of war is to patch over any honest perceptions of reality with lies dictated by the party line. His job is to tell you that you can't believe your own eyes, that shit is ice cream, and that you owe it to yourself to hurry back to the war with a positive attitude and slaughter people you don't even know, because if you don't, you're crazy' (*Phantom Blooper* 188–9). Hasford attacks all the system's props: careerist military, the brass, religion, military tradition, and, most of all, psychiatry. Apparently concerned with perceiving and clarifying troublesome underground matter so that emotional pain can be relieved, psychiatry becomes a mechanism that instead perverts one's sense of the world in order to reassert the military's definition of reality. For Hasford, it isn't just the lying but the murderousness of it all, the loss of personal understanding, the inversion of everything that seems right, so that one is convinced to do the wrong thing and run back to the killing. The inescapable Moebius strip of *Catch-22* has been replaced by a chain of overt lies. Eighteen-year-old combat veteran Bill Ehrhart, when questioned by his army chaplain about the reason he has fallen away from religion, reluctantly asks in return how it can be right to 'just say you're sorry and then go out and deliberately keep doin' the same stuff over and over again. You got no business tellin' guys like me it's okay to deal like that ... You've got an excuse, Colonel Glass has got an excuse, General Westmoreland has an excuse. President Johnson has an excuse ... Either you're a Christian, or you're not. There's nothing ambiguous about "Thou shalt not kill"' (*Vietnam-Perkasie* 276–7).[42] The priest knows he's beaten, that, in signing on to the military ticket, he has forfeited his right to talk about moral good: his job is to keep soldiers in the line. Ehrhart genuinely likes and feels sorry for the man but cannot continue to believe in a system that uses cultural forces civilians once trusted to support bankrupt policies. The sadly knowing eighteen-year-old Ehrhart is as disturbing as the psychotic rabbit hole down which Hasford's characters have disappeared. Ehrhart seeks and cannot find even the smallest measure of rectitude as his questions lead him directly up the chain of military and civilian leadership until he performs the forbidden: questioning Command.

By the end of the Second World War, American military psychiatry officially concluded that 'there is no such thing as "getting used to combat" ... psychiatric casualties are as inevitable as gunshot and shrapnel wounds in warfare ... Most men were ineffective after 180 or even 140 days' (quoted in Keegan *Face* 329). The post-war report specified that, at about ninety days, accustomed to his equipment, taking fire, making combat decisions despite panic attacks, the soldier would reach his peak ability. Even that ninety-day mark turned out to be optimistic. According to current thinking about the effect of increased battle 'tempo,' as it's called, the number of days a human can tolerate combat without psychiatric collapse or severe mental distress has been dropping. Psychologist, paratrooper, Army Ranger, Lieutenant-Colonel Dave Grossman, who trained soldiers for years and now trains police to 'bulletproof the mind,' concludes: 'We're placing people in combat for months on end and you know what we're finding? For those that did not otherwise become casualties, after 60 days of continuous combat 98% of all soldiers had become psychiatric casualties. The other 2% – they were sociopaths' (Grossman 2:1). Grossman's numbers are intriguing because they come from an Airborne officer and Ranger who is still actively engaged in training people to kill (his website, Killology.com, is fairly self-explanatory), who has had to deal with the realpolitik of military bureaucracy. If the maximum is sixty days, then the average point of breakdown will be less; indeed, given the army's propensity for pushing people beyond their limits, even that lower mark is probably an overestimate. Myths about American troops in the Second World War do not include the reality that most military actions were attended by thousands upon thousands of psychiatric casualties who, even if prevented from leaving the line (airmen in Bomber Command were denied all forms of stress leave), were useless or, worse, dangerous to themselves and each other in battle. Keeping men in the line didn't mean they could function.

In order to reduce anticipated psychiatric losses, the British Army introduced new training methods at the beginning of the Second World War, some of which are still in place (realistic obstacle courses, 'live-fire' exercises with real ammunition); others, like 'battle inoculation' (marching groups of men through slaughterhouses in order to desensitize them to the gore of industrial murder), have been discarded. Most of the methods share what psychologist B.F. Skinner calls 'operant conditioning': the deliberate training of animals (soldiers) to respond in particular ways to given stimuli (the noise and atmosphere of combat). Despite the slaughterhouse visits, what couldn't be inoculated against were the psychological effects of seeing high explosives partially or entirely destroy bodies. The most alert soldiers were aware of their fate well before landing, as the young Second Lieutenant Paul Fussell calculated: 'If this many were needed – a new class of almost one hundred graduated every week, and our class was the 321st – the rate of destruction was much greater than we'd ever allowed ourselves to contemplate ... We were meant to be expended, and that's why there were so many of us' (*Battle* 91). Fussell's realization that he is wastage, that some 32,100 men have already passed through the camp and been used up, only makes the speed and haphazard quality of the training worse: that he knows he isn't ready, and neither is his army.[43] The numbers are even more appalling because they refer to officers, each of whom is the leader of a larger number of equally poorly trained enlisted men. Fussell's understanding that only utmost chance will bring him through alive starts a lifetime of rage at the state. The same recognitions strike an uneasy Lewis Puller, Jr (son of Lieutenant-General Lewis 'Chesty' Puller, the most decorated and famous twentieth-century Marine) about twenty-five years later as he is rapidly pushed through Officer Candidate School on his way to Vietnam. After surviving a booby trap that costs him both legs at the pelvis, and most of both hands, Puller's partial body is returned to the United States. In a Veteran's Hospital for the first time, he cannot avoid the 'stark evidence that I was merely refuse to be discarded, and I was uncomfortable with the growing feeling that I had been used by my country. I was also apprehensive, to say the least, about the possibility of being warehoused at age twenty-three in an antiquated VA [Veteran's Administration] hospital' (Puller 253). Like Fussell, Puller has an inkling of what is coming – only a 'staggering attrition rate among young lieutenants' could create the 'frantic pace' of the training course (52). But those warnings don't help him with the reality of being dumped into Walter Reed Army Medical Center (the chief military hospital in the United States), or with the terror of being shut into a closet to suffocate: seeing wastage is bad enough, becoming it is horrendous. Soldiers lucky enough to survive the combat must confront their increasingly clear vision of what it is to be maimed or dead. Preparing for another day of flying, helicopter pilot Bob Mason seems to drift as he approaches the flight line with a new trainee pilot:

> I walked across the quarter mile of sand with Fisher. I kept checking my gear, like a novice. Pistol, flak vest, maps, chest protector. Oh, yeah, the chest protector is in the ship. Helmet. Courage. Where is my courage? On, yeah, my courage is in the ship.
>   'Lose something?' asked Fisher. He had been watching me check myself, patting my pockets and gear ... (Mason 460–1)

Recalled to the operation centre before he can crank the Huey, Mason discovers that the major has flown in his place: it's clear that Mason shouldn't fly. Repetitions once hard to learn but necessary for survival have become compulsive behaviours signifying a world of panic and superstition. It may not be quite accurate to say that his 'courage is in the ship,' but it isn't far off. The Huey he's spent so long flying is one of the few things in Vietnam he feels he has control over; it is a repository of hundreds of hours of expert piloting and battlefield decision making. The ship can renew Mason's confidence, enable and extend his self. But terror lives there too, and his pocket-slapping unfocused jitters make him visible to the military that will weed out even the most competent of soldiers. What happens to them is the same thing that happens to all weeds.

There can be no return to the thing the soldier most desires: pre-war existence. Not only may the external world have violently shifted (as it did for many veterans returning from Vietnam in the late 1960s), but the soldier's internal reality has so altered that home has vanished. Pre-war expectations about a universe that operates according to some hazy set of codes (bad things don't happen to people who wish others well; arbitrary death, maiming, loss of self cannot befall the indomitable hero) have been obliterated. Healing rites that worked to return warriors to the tribe can't help millions coming home from industrial slaughter.[44] In ritual combat, where each act has particular significance, return can be orchestrated: those same rituals, romantic myths to the contrary, will not function in the universe of high explosives, high technology, and wastage. The civilian walks through a pass to military territory: an infinite gateless wall blocks the return. No matter how much soldiers may loathe the military, they also take part of it away with them; perhaps only other soldiers can understand and welcome back veterans to the place that once was home. But salvaging the body comes first, as Dalton Trumbo predicted in his 1939 novel *Johnny Got His Gun* about Joe Bonham, a limbless, faceless First World War soldier rescued by contemporary medicine: 'The doctors were getting pretty smart especially now that they had had three or four years in the army with plenty of raw material to experiment on. If they got to you quickly enough so you didn't bleed to death they could save you from almost any kind of injury. Evidently they had got to him quickly enough' (Trumbo 82). Joe is doomed to finish his existence as a human stump of flesh, denied even the euthanasia he desires. He spends, or rather wastes, his time thinking of everything he can never do again. The machine of industrial warfare brings on a similarly enormous system of military medicine that capitalizes on the public flesh laid out before it. A badly wounded Joe Haldeman, who, unbeknownst to his surgeons, woke up during one of the operations that saved him, later gave one of his protagonists a vision from the operating table: 'Squinting into the bright light, the blood on [the surgeons'] green tunics could have been grease, the swathed bodies, odd soft machines that they were fixing' (*The Forever War* 173). The soldier is a body-machine, now separate from others.

The severely maimed Lewis Puller knows that he's been 'forever set apart from the rest of humanity' (186). Puller's will and youth make possible his survival through the years of surgeries and physical therapy that he must endure. He has been set apart and, more than that, set aside. His high-explosive rebirth into a partial human body is followed by surgical debridement, persistent detachings of veteran-tissue from the civilian self. Puller's anxiety becomes terror not only

at the loss of his legs but at the state of his shattered hands: 'I had
seen my right hand with its missing thumb and little finger earlier,
and I also knew that my left hand now retained only a thumb and half
a forefinger. The word *prehensile* no longer applied to me' (187).
Puller chooses his words with care. It is the small things that are
enormous – no longer part of the race with opposable thumbs, who
can speak to him as an equal? It isn't as if he has gained from the
exchange, being blasted into his new race of one. His partition cannot
be papered over by the goodwill of those around him. Much of his
survival he owes to his wife, Toddy, a woman who stands by him
apparently without fail or hesitation as he ascends to a new form of
life (there will be no medal for her, should she want it, nor medals for
any partners or families). As hurt soldier bodies pour off planes from
Afghanistan and Iraq (as of this writing, some 11,069 of them, with
another 1,502 dead, according to U.S. Central Command – and
somewhere between 20,000 and 100,000 Iraqis killed, the first number from the Pentagon, the second from a tightly scrutinized article
published in *The Lancet*), they too will ponder the deficits they
incurred, the dreadful moments, less than seconds, in which explosions bankrupted their military worth:

> 'I think I should be dead right now,' the 20-year-old specialist
> Acosta said ... 'But I feel like I failed myself. If I hadn't dropped [the
> grenade], I would still have my hand.'
> Reminded that he saved his friend's life, specialist Acosta stared
> straight ahead and remained silent. (Banerjee 'Rebuilding' 1)

Heroic acts cannot erase the eternal moment of exile, the point at
which the soldier forever left behind an existence perhaps taken for
granted. At each moment the grenade drops, and then drops again, a
persistent dreadful past always present. What the whole soldier has
done out of generosity (absorb the blast), the amputee must struggle
with, regret, try to hold on to with the missing hand.

There's always more when it comes to doing pain mathematics.
Body armour functions to preserve lives but also to channel harm.
The result is 'a stream of young soldiers with wounds so devastating
that they probably would have been fatal in any previous war ...

Wounds involve severe damage to the head and eyes – injuries that leave soldiers brain damaged or blind, or both' (Vick). Previously, blasts of such force would have ripped open the torso or caused a fatal amount of trauma: not any more ('evidently they had got to him quickly enough' [Trumbo 82]). Doctors at forward hospitals now face, and create, ethical problems. One combat surgeon, Lieutenant-Colonel Robert Carroll, observes bleakly: 'We're saving more people than should be saved, probably' (Vick). Surgery, then ethics. Medicine is up to the task of preventing outright wastage, but the results will be set on the stoop for the victims' families to collect and cope with for the rest of the survivors' lives. Those who have lost brain material (craniotomies to remove dead sections of brain occur at the rate of about one a day) may or may not know what they've given in combat. The doctors don't know what the soldiers will understand. Witness to a parade of bodies forever severed from wholeness, Puller voices his dismay to his fellow officer amputees, only to be told 'to get down to the enlisted wards to see the real spoils of war' (211). 'Spoils' carries the same sickening chill as do 'waste' and 'invalid': bodies left out in the steel rain have spoiled, are rotten meat. There's a lot more rotting in the lower ranks simply because there's more to discard. Lieutenant-Colonel Carroll's reflection about the sheer number who have been saved (whether they should be or not) is a factor of his position as the hand that lifts out the military's dross, cuts away dead flesh so the rest of the body can continue fighting. Carroll is at the transition point where the pushing military hand forces the grasping medical one to catch the body. Those once-human will now be, as Puller says, 'set apart.'

The veteran calculates a variety of betrayals: betrayal of the body, of the past, of what the country once was (or was perceived to be), of trust.[45] The possibility that Puller will be forced into a dark room until he has the good grace to die is prevented only by his wife's personal strength. Paul Fussell looks stonily at the badly trained American army fighting the ground war in Europe and concludes that it was an 'unintended form of eugenics, clearing the population of the dumbest, the least skilled, the least promising of all young American males,' whose fate 'constituted an unintended but inescapable holocaust' (*Battle* 172). In his memoir and in *Wartime* (1989) Fussell attacks the

fantasy of the 'good war' created by the American myth machine
(Mary O'Hare calls John Wayne and Frank Sinatra 'glamorous, war-
loving, dirty old men,' in Kurt Vonnegut's *Slaughterhouse-Five* [14]).
Instead of displaying battlefield ability, the infantry's task was to tire
out the Wehrmacht; if the army was largely wasted, at least the post-
war employment force would be cleared for the booming generations
to follow. One is perhaps more familiar with this suggestion in the
context of the First World War, a war with few perceived winners and
millions of working-class losers, than in regard to the Second World
War. But the undeclared wars in Korea, Vietnam, Afghanistan, and the
Persian Gulf have severe implications for the working poor and the
under-educated of the United States, not only in the cost to bodies,
minds, and families but in the profound war deficits that will attack
the working class most heavily. Both Puller and Fussell savage the
callousness and idiocy with which the state expends the lives of its citizens, who, while they may at first believe what they've been told,
begin to question their government when they are wounded physically
or emotionally (or both). Even the once-true believer William Ehrhart
concludes on reading *The Pentagon Papers* (1971) that, when it came
to Johnson, Nixon, McNamara, and Westmoreland, he had 'been a
fool, ignorant and naïve. A sucker. For such men, I had become a mur-
derer. For such men, I had forfeited my honor, my self-respect, and my
humanity. For such men, I had been willing to lay down my life. And
I had been nothing more to them than a hired gun, a triggerman, a
stooge, a tool to be used and discarded, an insignificant statistic'
(Ehrhart *Passing Time* 175). Ehrhart's catalogue parallels Puller's
comments about the 'true spoils of war'; the proof of wastage is to
know that you've been conned, have killed for other killers, have
given up innocence for immoral power-brokers, have forfeited the
interior being for the exterior myth machine that uses humans for the
shallowest of reasons, or so as not to be caught in a lie. It is impossible
to equate Ehrhart's moral wounding with Puller's bodily loss, but the
soldiers are twinned in outrage at becoming discards, statistics.

It seems at times impossible that the self could withstand the realization that states do not care about the citizen, that states will
blithely expend citizens' lives, seeing in them a flesh bridge to its
chosen destinations. Fussell, who took part in what was first ironi-

cally, and then flatly, called 'the good war,' is stunned by what he sees as the re-creation of the German war state in the Bundeswehr, at the fact that '90 percent of American assertions about other countries [turned out] to be cunning lies. It seemed I'd been demeaned and all but killed fighting the wrong enemy' (*Battle* 208). Such comments are no different from those of other veterans who, on looking back at state policy, know themselves to have been, and still be, taken for dupes. Physical hurts will be handled by warehousing the maimed, moral damage will be ignored by policy makers on their way to the next war.[46]

Perhaps there are degrees of betrayal. The body can be broken without being utterly destroyed, a human can be reduced yet not cease functioning or lose all worth. The citizen has, after all, agreed to a trade-off. The state proposes to use the military body for the good of the whole: between itself and the enemy the state may need to interpose flesh cushions made up of those who are perceived as worth less than the entire, certainly worth less than the military and civilian leadership. The state sorts out bodies according to their value, the power inherent in them, their connections to other bodies of significance, wealth, knowledge – and the citizen submits. But no citizen sets out to be, as Ehrhart laments, 'a sucker.' Even, in fact especially, after paying as much as they have, veterans like Puller and Ehrhart are ardent about the republic that has used them as gun wadding. Equally caught in the rifle bore are their betrayed families. Joseph Heller's brutal form letter to grieving relations shows how the individual is valued: '*Dear Mrs., Mr., Miss, or Mr. and Mrs. Daneeka: Words cannot express the deep personal grief I experienced when your husband, son, father or brother was killed, wounded or reported missing in action*' (427). In the end, no military really separates the victims. As Vietnam veteran Philip Beidler warns grimly about the 'Hoo-ah kids,' the United States prepares for the next round of combat: 'Go out and do the mission, we'll say. Deal with the Charlies or the Sammies or the ragheads or whatever. From where I stand, after all these years, I can offer only one sure lesson: Don't come home remotely expecting anyone to care' (*Late Night* 212).[47] And yet. Surely there are veterans whose lives have been greatly enriched by war, who have learned the value of peace and democracy. Surely for some, when the war is over, it's over, and what follows is an era of harmony. Surely not *all* veterans were wounded or maimed, physically or emotionally, but were helped back to civil society by a grateful nation.

But, in truth, after war there is more war – war with military and medical bureaucracies. In no shape to argue for themselves against a state full of lawyers, damaged veterans must be ready to fight for help: Puller finds it 'remarkable ... that my country was so vast that it could simply swallow up the dead and maimed from my generation's war and continue to march as if nothing was amiss' (301). The physically immense country is overshadowed by a larger indifference to, or ignorance of, war costs – gargantuan refusals to deal with the end products of the war (excluding overt moral victories, reparation payments, land deals). In a 2004 press conference, when asked about the ban on photographs of soldiers' caskets being returned to Dover Air Force Base, President George W. Bush said: 'Look, nobody likes to see dead people on their television screens – I don't. It's a tough time for the American people to see that. It's gut-wrenching' (Bush). Lewis Carroll's Queen of Heart's logic operates here: since it is difficult to see, then it can't be shown, which will make it even more difficult to see. Some war products – action shots, technology on the move, medical examinations of the captured enemy – these are for the public. But death and maiming are private, ideally so private that we never know they occur. Joseph Heller knows: Nurse Duckett warns that 'they're going to disappear [Dunbar],' one of Yossarian's friends. A baffled Yossarian retorts: 'It doesn't make sense. It isn't even good grammar. What the hell does it mean when they disappear somebody?' (Heller 454). Contrariwise, the grammar is perfect. No one knows what 'being disappeared' means until it happens to them, and then it's too late. The state controls the language, the pictures, and has been busy rendering Puller's fellow combatants invisible. Impressive as the black granite wall is, it offers signifiers only; better if the prospective visitor were forced to walk through hallways of live paraplegic veterans before reaching the list of the dead (Alice realizes: 'You're nothing but a pack of cards!').

The war for survival in peacetime puts the wasted soldier, useless now for the next technological cycle, on the defensive; Henry Ford considered but then discarded the idea of training thousands of First World War veteran single and double amputees as mechanics who would be better able to work at specific tasks than humans with complete bodies (Seltzer 157). Systems that call out to other expensive systems have no traffic with broken bodies, nor will the economy be directed to a two-front fight (medicine *and* defence spending). Looking back at military pension costs and facing a wider disaster in 1939, Britain 'determined not to incur such a bill again [and to] ... lock the medical policy machine into a very restrictive definition of war neurosis' (Shephard 165). If there's no language for neurosis, then there can be no pensions for it either (just as if there are no pictures of caskets at Dover, there are no dead). Phrases like 'shell shock' and 'battle fatigue' (later 'Post-traumatic Stress Disorder' [PTSD]) were not only too prevalent for the military's liking but were nebulous terms attached to virtually any irruption of anxiety or nervousness. Wary of the as-yet unknown effects of long-range bombing and air war, the military sought to purge psychiatric language from its lexicon. Denial is common to militaries that wish to fight new battles without having to cover the after-action costs of old ones. After war, the veteran faces the task of forcing the state to speak about the unspeakable (there will be no language of illness, no mention of pensions, no visible support for war-related problems); the veteran must persist in remembering, in recalling for the state and the public what has occurred. The fight is on for the material that will compose the collective memory, for a new set of signifiers, for visibility, for new citizen-soldier organizations, for a definition of patriot that includes angry, wounded veterans.

It should be little surprise, then, to find that 'as of July 1995, VA had denied almost 95 percent of the 4,144 claims it had processed for Persian Gulf veterans claiming such disabilities' (United States General Accounting Office *Veterans' Compensation* 1). The General Accounting Office (GAO), a powerful oversight body, was called in to examine how the VA could turn the wheel so quickly on veterans of such a public, apparently successful, war.[48] There were two main issues: first, it seemed that only the Reservists got sick or applied for pensions (yet they had served theatre-wide during the war); second, the denial rate had nothing to do with the veterans' illnesses (they were demonstrably sick). The GAO determined the answers to these problems, as did investigative reporter Seymour Hersh. They found that the Reservists, because they had civilian jobs, weren't afraid to make health claims against the VA, whereas career officers and troops didn't make claims because President Bill Clinton had promised to reduce the army – unstated was that the first to go would be troops complaining of health problems. The reason the claims were denied in such quantity was that most veterans didn't get sick immediately on return from the war zone, and claims made after two years could automatically be denied using 1991 VA policy. It just so happened that the fuse on what came to be called Gulf War Illness (GWI; sometimes GWS, Gulf War Syndrome) took slightly more than two years to burn down; then thousands, and finally hundreds of thousands, began reporting sicknesses. By April 2002, what had started as a few red spots on the collective military skin had become a debilitating rash: '7,758 Desert Storm vets have died, while 198,716 vets have filed claims for medical and compensation benefits ... The 198,716 figure represents a staggering 28 percent of the 696,579 vets who fought in the Gulf War conflict' (Hackworth 'Oops').[49] The numbers come from the VA and the commentary from life-long professional soldier Colonel David Hackworth (one of the few career military to declare publicly in 1971, that the Vietnam War was unwinnable). Even without words, the veterans still got sick and came to collect damages. Silence is key to the Revolution in Military Affairs, and an eerie quiet about human health pervades the RMA warfare sphere. Thousands of ill veterans slammed up against battlespace's glass ceiling that excludes the failure they represent. As Richard J. Danzig, secretary of the Navy between 1998 and 2001, told Seymour Hersh: 'The senior military people say, "What we do is fight. This other stuff is not my arena. It's messy. Sickness is not what I deal with"' (*Against* 43). With their slowly developing signs, the uncertainty of cure, apparently endless emotional and physical woes, the sick created profound messiness in a world driven by PowerPoint slides and textureless computer graphics of battlespace. The RMA war body is not physically constituted for the ill, who interrupt battle flow and combat tempo. There is no humanspace in battlespace – that, after all, is the point.

Humanspace includes the reality of lying in the elephant grass of Vietnam, watching liquid defoliant drift down from circling planes, getting sick, and staying sick. Agent Orange, a fifty-fifty mixture of chemicals 2,4-D, and 2,4,5-T (both laced with dioxin and banned for use in the United States), was showered onto Vietnam in the belief that destroying the triple-canopy jungle would halt North Vietnamese military successes; when Operation Ranch Hand, responsible for spraying the country with defoliants, was complete, some nineteen million gallons had been dropped on the country, the Americans fighting there, and Vietnamese soldiers and civilians alike (Lewis Publishing Co.). Dioxin is a substance whose toxicity is measured in parts per million, not millions of gallons.

Within a year of returning from Vietnam, those veterans who were exposed to Agents Orange, Blue, and White (originally created as chemical weapons by Dow Chemical, Monsanto, Uniroyal, and named for the particular colour of the stripe painted on the chemical drums) found themselves sick or dying, their wives miscarrying, their newly conceived children deformed or stillborn. The military claimed that no personnel had been allowed into defoliated areas until six weeks had passed, to which one soldier retorted: 'That's ridiculous. We would be dropped off at an ambush site and spend hours actually lying in the stuff. It would actually be comin' down on us' (Wilcox *Waiting* 39). The men doing the spraying didn't wear chemical gear or masks, nor were they ever warned about the herbicides' toxicity.[50] Agents Orange-, Blue-, and White-related illnesses, as documented by the prestigious National Academy of Sciences, include 'chronic lymphocytic leukemia, soft-tissue sarcoma, non-Hodgkin's lymphoma, Hodgkin's disease, Chloracne' (*Veterans and Agent Orange: Update 2002* 8), to say nothing of the stillborn babies and live births with severe birth defects (spina bifida, missing brain tissue, missing eyes or internal organs, fingerless hands). In the 1970s the Vietnam vets who recalled being sprayed or who had worked on Ranch Hand came together in a class-action suit that would last over a decade, beginning a legal fight between human wastage and the defence establishment.

The military denial that defoliants capable of eradicating triple-canopy jungle had ill effects on the health of the thousands of sol-

UC-123 Aircraft (USAF)
DISSEMINATION OF HERBICIDES
OPERATION RANCH HAND
SOUTH VIETNAM, 1967

20 **Chemical Solutions**: Air Force C-123 tankers spraying Agent Orange in Operation Ranch Hand. Facing Vietnamese triple-canopy jungle, Ranch Hand worked to eliminate the enemy's cover, and the enemy. Inadvertently, it eliminated a good deal of the American Army as well and took a predictable toll on all Vietnamese in the affected areas (which is to say, most of the country).

diers, doctors, nurses (at forward surgical hospitals) was another of
the war's big lies. Eventually the class-action (or mass tort, where
'tort' is a personal-injury case) suit came to represent 2.4 million
veterans and their relatives, naming Dow Chemical, Monsanto, and
Uniroyal, among others, as defendants. Dow in turn sued the Department of Defense, arguing that Dow 'had sold its products to the
Department of Defense in good faith and from then on it was really
the government's responsibility to use it properly' (Wilcox *Waiting*
104). After five years and hundreds of thousand of legal dollars, the
veterans and the chemical companies settled out of court. The soldiers and the other class-action participants (wives who had miscar-
ried, family members fighting on behalf of their dead relatives or
malformed children) divided almost perfectly in two: the stronger
group argued for taking the money; the weaker for full disclosure
from Dow about Agent Orange's toxicity, along with something
rather old-fashioned: an apology. The financial settlement proved to
be additional misery because, in order to accept payout, a veteran
would have to waive his or her rights to Veterans' Administration
support – it could cost more to collect from Dow than not. One
defeated veteran deduced that 'they're just waiting for all of us to die,
and then someone can say, "Oh dear, maybe we did make a mistake
with this Agent Orange." At the rate things are going they won't have
to wait very long' (Wilcox *Waiting* 15). As of this writing, the veterans' claims have been largely if grudgingly substantiated by the
National Academy of Sciences. It is true that it is easier for exposed
veterans to make successful claims. It is also true that, as of today,
many of the Agent Orange veterans have died.

Agent Orange became a directional sign for the veterans, the military, and the military-industrial complex. Veterans saw in the lies,
illness, and genetic damage the whole of the Vietnam War, itself a
cultural sign for lies, disfigurement, and death. The military, still
reeling from losing a war upon which had been lavished every technological excess available, learned that it could also lose the peace of
defeat. The military-industrial complex already knew not to get
caught again, a moral that would prove crucial as the next wave of
sick veterans came home from the 1991 Gulf War. Because the first
Gulf War performed military (acting as a test-bed for new materiel

and revised manoeuvre-warfare strategy), political (uniting world forces to free Kuwait), and cultural work (putting paid to 'Vietnam syndrome,' that is, 'losing') for the United States, it could not be tarnished. There would be no Agent Orange in the Gulf, determined the military contractors and the Department of Defense. And so, when Desert Storm veterans, combatants and non-combatants alike, began to complain of a constellation of illnesses (joint pain, dizziness, nausea, persistent rashes, headaches, memory loss, chronic fatigue), they were met with disbelief, hostility, and finally, as the VA began to see thousands of cases, Prozac.

Many Vietnam veterans had handled their trauma differently from previous generations of soldiers: they talked openly to each other about their experiences, participating in talking cures ('rap groups') under the guidance of psychiatrists Robert Jay Lifton and Chaim Shatan. Storefront psychiatric clinics were set up across the United States so that veterans could be treated in their own towns. Out of such populist medicine came the increasingly technical language of Post-Traumatic Stress Disorder, formalized in the 1980 *Diagnostic and Statistical Manual of Mental Disorders, 3rd Edition* (DSM-III), the medical profession's codex produced by the American Psychiatric Association.

Assuming that the collections of symptoms reported by Gulf War soldiers were stress-related, psychiatrists used to dealing with Vietnam veterans handed out newly created anti-depressants like Prozac (or other Selective Serotonin Reuptake Inhibitors, SSRIs). Whereas during and after the Vietnam War the veterans fought for psychiatric care, now the Gulf soldiers struggled to prove that their problems were physical, not psychiatric. Desperate to protect its recent victory, the Pentagon dug in against its own veterans and hired the RAND Corporation, a non-profit think tank created by the military-industrial complex in 1948, to assess Gulf War Illness. The RAND's eight-volume study (published between 1999 and 2001) examines each part of the illness, from psychological stress to oil-well fire pollutants, to the anti-nerve gas pill pyridostigmine bromide (PB). At the close of one volume, the author lashes out at what he calls 'the subculture of "Gulf War Syndrome,"' arguing that 'the rapidity with which an illness narrative may spread, be shared, and be responsible for shared symptoms has been demonstrated in a number of studies of so-called hysterical epidemics in school and office buildings' (Marlowe 164). That summation, backed by the full force of the RAND and by extension the Pentagon, attacks ill veterans in a variety of ways: it suggests that they are not steadfast military but rather hysterically suggestible civilian workers; that they share their illness around as do malingerers (they are not the culture but a subculture – with all the import of that which is lower, lesser, unworthy); that the illness is in fact a story told from one shirker to another; that it has no basis in science, unlike the science that has documented surges of hysteria. The RAND reports, together with two other parallel RAND studies of GWI, represent more than 1,800 printed pages of official story about what might be wrong with the veterans.

The amount and ferocity of veteran complaints over the course of more than ten years have forced the Pentagon to admit, finally, that *something* is wrong. There are two chief contestants for the role of toxic bad guy in Desert Storm. The first is an anti-Soman (a nerve agent that causes death by asphyxiation within a minute of inhalation) drug, pyridostigmine bromide. PB was rushed through clinical trials and given to the troops as they arrived in Saudi Arabia, the staging ground for the invasion of Iraq. The case for PB as the villain is strong, although one common argument discounting PB's toxicity is that, because the drug is tolerated in large doses by patients with myasthenia gravis (a potentially fatal autoimmune disease that prevents nerves from receiving signals, resulting in debilitating weakness and the attendant loss of the ability to breathe unassisted), healthy patients will be untroubled by it. But even the RAND had to admit that justifying PB this way was like 'assuming that since high doses of insulin are tolerated – or even necessary – in some patients with diabetes ... [that] a smaller dose of insulin should surely be safe in those without diabetes' (Colomb xxiv). Seymour Hersh notes that, when given to healthy people, PB can result in 'a neurological response known as bromide intoxication, the symptoms of which included confusion, tremor, memory loss, stupor, and coma,' almost all symptoms reported by the veterans (*Against* 22). While the troops were supposed to be given the choice of whether or not they wished to take PB, the reality seemed otherwise. Only in the second Gulf

War have the options been clarified, with the result that some soldiers have refused PB (Anthony Swofford recalls being informed by his staff sergeant: 'Don't fucking ask me what it [PB] is. I'm taking it too. Do you want to fucking live or do you want to fucking die?' – an exhortation that resulted in the men taking three times the suggested dose [Swofford 182]). PB began to look like the author of the strange, debilitating blur of chronic pain, rashes, fatigue, memory disorders collected under the Gulf War Illness umbrella. Even more alarming was the newly discovered 'ability of PB to cross the blood-brain barrier under circumstances of stress' (Colomb 73). As it turned out, stress, heat, and multiple chemical interaction (PB plus DEET [bug spray]), all of which were available in quantity through-out Desert Storm, allowed PB to move across the blood-brain barrier, opening up the central nervous system to attack. It was a bitter pill for the Pentagon to swallow: the RAND report confirmed that not only would PB be useless in a Soman attack but that the drug had its own significant, toxic after-effects.

Why carry on looking for the chemical criminal when PB had already been handcuffed and charged? One could say that, for the nearly 200,000 veterans who have applied for assistance in the last thirteen years, it was bad luck for them to have adverse reactions to PB but at least they didn't encounter Soman (which would surely have killed them) and, even if the quality of their lives has been sharply diminished, they did survive. True, they are wastage, but they have all their limbs.

Recall the Agent Orange struggle however, and pause. Citizens of Iraq weren't being ordered into formation three times a day and forced to take expensive pyridostigmine bromide pills. In fact, they had no access to the drug at all. Just as Agent Orange sickened com-bat veterans, their families, Vietnam and its people (all suffering high rates of soft-tissue sarcomas and other dioxin-induced illness), so too was post-war Iraq, the ground and the people, under attack. An Iraqi doctor, a Fellow of Britain's Royal College of Physicians and Sur-geons, put it this way: '"Before the Gulf war, we had only three or four deaths in a month from cancer ... Now it's 30 to 35 patients dying every month, and that's just in my department. That is 12 times the increase in cancer mortality. Our studies indicate 40% to 48% of the population in this area will get cancer – in five years' time to begin with"' (Pilger). What was it that was on the ground, available to the whole population, that could cause a cancer explosion? What accounted not only for dizziness and memory loss in American veterans but also unusual cancer rates in the Iraqi population?

Iraqi scientists and American veterans turned away from PB and towards a second possible suspect: depleted uranium (DU). DU, already the star of a chapter in this narrative, the centrepiece of the Armour Piercing, Fin Stabilized, Discarding Sabot round fired from the Abrams M1A2 main battle tank; DU, enriched uranium's dense waste-product, given free to arms manufacturers in the United States; DU, the pyrophoric centre of 30 mm slugs fired from the A-10 Warthog aircraft at a rate of 3,900 rounds per minute, the same slugs that, shattering on hard targets, vapourize into radioactive fumes and pyrophoric dust – this was the new culprit. What quantity of such ammunition had been fired in a few days of war? How much DU could really be lying in the sands, leaching into the water table, and moving into the food chain, of Iraq?

The Pentagon, the Federation of American Scientists, and veterans' groups agreed on the amount: at least 315 tons of radioactive American DU were dropped on Iraq (some estimates go as high as 350 tons). The tonnage came from more than 14,000 tank sabots (half lost in practice shooting, a quarter in combat, another quarter burned in a fire) and at least 783,000 30 mm rounds fired from the A-10 Warthog's gatlings. Another 127 tons (this number is known to be low because the Pentagon has so far refused to release the true DU expenditure) have been fired on Iraq during the second Gulf War (Gonzalez). A primary air-to-ground attack aircraft, the Warthog also sprayed some twelve tons of DU shells on the Balkans during the Kosovo air war, and somewhat less on Somalia. A DU cloud can easily be carried thirty kilometres downrange by the wind; the fine dust mixes with earth used for food crops, migrating to grazing animals' milk. The Pentagon's repeated claims that DU is harmless are put into perspective by the 1979 closure of New York State's National Lead Industries Plant, which, when it vented a quantity of DU equal to that found in one single 30 mm Warthog shell, was permanently shut down by the Environmental Protection Agency (Gonzalez). In

1991 the United Kingdom Atomic Energy Authority warned that the then-reported forty tons of radioactive DU left behind by coalition forces had to be cleaned up: the real numbers (almost eight times that amount) spoke of an enormous environmental disaster. In April 2003 repeated misquoting of a Royal Society report on DU forced a corrective media release by the Society explaining that it had never given DU a clean bill of health and calling on the American government to commit itself to removing radioactive dust from Iraq's soil (Royal Society).

The Pentagon stands by its claims that DU is simply a dense, cheap material that makes for excellent munitions and poses no threat to human health (unless DU splinters lodge in the body, an event that occurred to over thirty Desert Storm friendly-fire survivors – they've been added to the list of so-called atomic vets, the 382,000 American soldiers repeatedly and deliberately exposed to atomic bomb tests from the 1950s to the early 1960s). The RAND's DU text (volume 7 of the Gulf War Illness series) not only washes clean the Pentagon, the military, the atomic industry, and the consciences of all those involved but, most of all, exonerates the material itself. What the RAND report omits is the care taken with handling DU on the continental United States. At the Aberdeen proving grounds in Maryland, DU shells are tested in a piece of equipment called a 'Superbox' (fig. 21). Sabots are fired down the Superbox's alley, doors slam shut fifty-six milliseconds after the round has entered the tunnel, and, when the round hits the target and shatters, the radioactive DU vapour is immediately drawn off, filtered, and contained by ventilation stacks. The whole trap operates with perfect timing, else the apparently harmless DU materials would be released into the (American) civilian environment. So incensed has the RAND (the supposed arm's length – in itself an intriguing phrase – scientific establishment) been at the treatment of DU that it published Undersecretary of Defense for Personnel and Readiness Bernard Rostker's *Depleted Uranium: A Case Study of Good and Evil* (2002), which attacks the media for spreading ostensibly erroneous information about DU. Rostker's deliberately chosen Manichean title positions the government and RAND's reports as Truth and those provided by veterans, concerned doctors (Dr Helen Caldicott has asserted that using DU is

21   **Supersafe**: The DU 'Superbox' at the Aberdeen proving grounds (Maryland) is the only safe way to handle DU munitions. The shells are fired down the long forward tunnel into the central chamber, where they hit their targets. The piping on the right draws off and contains radioactive vapours and dust caused by the rounds' impact.

equivalent to fighting an undeclared nuclear war), reporters, and anti-war activists as 'junk science.' To this moment, the state, the military, and the military-medical and weapons establishments stand by the assertion that because 'there are no peer-reviewed published reports of detectable increases of cancer or other negative health effects from radiation exposure to inhaled or ingested natural uranium at levels far exceeding those likely in the Gulf,' DU is harmless (Harley et al. RAND xviii). The American military world has put the facts about DU into its Superbox where the radioactive news is siphoned off and scrubbed clean.

In 1994 the U.S. Army Environmental Policy Institute issued a report to Congress in which it gave depleted uranium one of its many clean bills of health, adding that no teratogenic effects had been seen in veterans' newborn children, and more: 'It is important to note, however, that no female soldiers were involved in the friendly fire incidents and none served on the recovery and maintenance teams' (U.S. Army Environmental Policy Institute). This report somehow missed not only Carol Picou, a licensed nurse with fifteen years in the army, but another seven women who volunteered to recover bodies from the infamous 'Highway of Death' where hundreds of vehicles fleeing the American invasion were trapped and cut to pieces by DU-firing tanks and planes. The 'Highway of Death' was littered with charred vehicles emitting radioactively dirty DU smoke, dust, and vapour. Picou went forward with 150 soldiers and became ill immediately; within a year she reported that 'My muscles have deteriorated. I have no control over my bowels or my bladder at all anymore. The army issued me diapers and said that I could catheterize myself for the rest of my life. I've been catheterizing myself since January of 1992. I've been wearing diapers for the same time' (Picou 46). In addition, Picou has been diagnosed with 'suspicious squamous cancerous cells of the uterus' (46). She has received no diagnosis. Of Picou's forward group, forty are sick, six have died. Another 150 stayed behind and have not reported any illness.

I have said that DU is a technological battlespace hero and that pyridostigmine bromide (PB) turned out to be a failure. The RAND concludes that 'the use of DU munitions and armour is likely to expand greatly over the coming years, both in the U.S. military and in other countries. It is therefore important to continue research to further our knowledge of any potential health risks that might result from different levels and pathways of exposure' (Harley et al. 72). Because the United States has been supplying its current allies with DU, because DU is now widely available on the world market, and because the United States has a ready and free supply of this reactor junk by-product, DU will be protected and disseminated and PB will be discarded. Vilifying PB allows the military to admit to a relatively smaller error while covering up its true desire: to use depleted uranium where, when, in the quantity, and on whom it likes.

Beneath the survivor's skin pounds the hammer of military practice. Those who escaped apparently unscathed from the Second

World War, who returned to an era of material prosperity unequalled in width and length at any other time or place on the planet, have since had to confront the once covered past. The war machine goes on cranking, vibrating through the days and nights, waiting for a chance to return to full operation (retired veterans who thought themselves free of war may suddenly begin to suffer night terrors, free-floating panic attacks, unexpected phobic responses to the world). After the First World War, Robert Graves recalled being 'mentally and nervously organized for war,' his room full of exploding shells; that 'strangers in daytime would assume the faces of friends who had been killed'; and when he came to 'revisit my favourite country, I could not help seeing it as a prospective battlefield' (*Good-bye* 235). There is little evidence that such habits fade with time. War has been instructive, has installed itself so as never to be evicted. Life is a war zone, the dead return, the most pleasant walks must be scrutinized: there's no choice about it. Even in his late eighties, Graves, the ardent pacifist who excoriated militarism, would say of his continuing misery: 'Well, you know, I killed too many people' (*Dear Robert* xvii). Apparently, his life's work of reflection and penance didn't lift his sorrow or return his civilian identity. More than fifty years after the Second World War, Paul Fussell still finds himself at the mercy of loud noises, still 'button[ing] all pocket flaps religiously. I still count silently to accompany such repeated body movements as putting on trousers or walking at the official rate of 120 steps per minute ... And when riding in a car or on a train, I still locate in the passing landscape good positions for machine guns, antitank guns, or minefields' (*Battle* 183). The language of the perpetual is the military hammer striking time, setting the cadence for efficiently performed quotidian acts. The war machine internalizes the mechanisms of life, creating repetitive gestures that are apparent in the language different soldiers use to tell similar stories. Another Second World War veteran says: 'I'd be driving home from work, and I'd look at the hills, and I'd say, "What a fortification. They could have gun positions there, and I could be ambushed going down the road. I'm wide open here"' (Takiff 257). Fields of imaginary fire and grid coordinates persist: being caught in the 'killbox' is the waking nightmare. Stereoscopic war vision allows the soldier limited mastery over and admiration of the land. All the veterans continue to plan militarily for the troops, envisioning machine and anti-tank guns, minefields in place; none speaks of retreat or digging foxholes. All landscapes are battlespaces now, all motions are military ones. The soldiers know this and work to alter the mechanisms, to recognize 'the soldier in us – the soldier that we were' and to 'shrink that soldier.' Otherwise, they will be fated to live 'always on the move. Securing the perimeter' (Takiff 431).

Even shrinking the soldier may not be enough. Fortifications that have held for years can collapse: 'Working everyday, from 1945 until 1995, I was so tired I didn't have time to let what happened overseas bother me. I could push it aside. But from '95 on, I just can't push it aside anymore' (Takiff 494). The past rushes in with Second World War veterans' silent memories. Vietnam veterans were more likely to talk, partly because of cultural changes regarding psychiatry, partly because the undeclared war became a miserable, protracted loss. But even the Vietnam veterans find that their work may not have lifted the war weight much: 'I can tell you it's never more than a heartbeat away,' says one fearful soldier (Takiff 521). Anthony Swofford, Desert Storm Marine sniper and literary heir of Graves, Fussell, Herr, and Hasford, dreams of 'coughing up pieces of shattered glass [from the rifle's telescopic sight], but no blood issues from my mouth ... I know my belly is stuffed with glass and that it might take me years to expel all of it' (Swofford 125). The nightmare of seeing death coming (being shot through a rifle scope as one looks down it) becomes a story of pain and time. The veteran is full of now purposeless, smashed machinery that must be removed – provided it doesn't cut him open in the process (Graves and Fussell's visions of battlefields have been transformed into blindness, illness). Perhaps even more worrisome is the apparent lack of effect, the bloodless expectoration; the body no longer works as it once did – it is bloodless (Robert Jay Lifton might find in it an example of 'psychic numbing,' the loss of the ability to feel). One of Swofford's friends, apparently suffering full-blown paranoid behaviour, reminds him: 'We fired the same rifle. You have the same problems as me' (118). The soldier has been part of that 'band of brothers' about which Shakespeare's Henry V spoke so desperately on St Crispin's day, but he now must live for himself, separated. He lets go of the rifle and the power that went with it. Fear, which the soldier expected to end once off the battle-field, continues, even worsens. How *much* glass, exactly, does Swof-ford have inside his belly? Is the supply renewed each day? Is the band of brothers there to help him spit it out each night? Certainly, there is no easy regurgitation.

Presumably Swofford's book is part of the glass-expulsion pro-cess, as it has been for many soldier authors (Robert Graves, William Golding, Norman Mailer, Joseph Heller, Philip Caputo, Joe Halde-man, Tim O'Brien). The act of writing guarantees neither that the work is, or ever will be, complete, nor that the nights won't be as full of war as ever. Lewis Puller's post-war unsuccessful run for Con-gress resulted in a long black drop into depression and more than a decade of alcoholism. After struggling with the diseases, drying out and handling his life, Puller hoped that writing *Fortunate Son* (1992) would allow him to 'forgive my government, to forgive those whose views and actions concerning the war differed from mine, and to forgive myself,' and in the process 'move into the present, attain a degree of serenity and find the reason for which I had been spared' (429). His success in overcoming alcoholism, practising law, raising a family, finishing his gruellingly honest memoir was matched by his winning the Pulitzer Prize and in the process reminding the United States that, one year after the first Gulf War, a Vietnam veteran had been forgotten and was still suffering through each day. The honours and praise that fell on Puller didn't prevent his alcoholic relapse or the departure of his beloved wife. They also didn't prevent his sui-cide in 1994.

Dave Grossman would perhaps say that Puller was insufficiently prepared for war. Grossman travels and lectures three hundred days of the year to 'bulletproof' (desensitize) police officers and military pro-fessionals who expect to use 'deadly force' (kill someone) in the course of their careers. Grossman distinguishes murder and atrocity from what he calls a 'righteous kill' (where the officer or soldier is on legal and moral grounds to commit murder). In a righteous kill 'there is not necessarily any backlash of remorse and nausea – among trained qualified officers – they feel that satisfaction of a job well done, and then what happens to them is that they feel a little bad that they don't feel bad ... They have already bulletproofed the mind'

(2:1). One way to avoid 'shrinking the soldier,' as one veteran put it, is to 'bulletproof the mind,' to accept Grossman's world-view that there are three kinds of people: wolves, sheep, and sheepdogs. The sheepdogs (soldiers, police officers) protect the sheep (civilians) from wolves (bad guys). It's even simpler in war, for Grossman: 'There are only two kinds of individuals on the battlefield – warriors and victims. Decide now which one you are' (1:1). What Grossman's false dilemma offers is a chance for people to have wars for as long as they like, without suffering any nagging psychological side-effects. War will follow war – forever – because kills can be made righteously. One of Joe Haldeman's future soldiers explains the clear and natural pattern of his life: he was 'born, raised and drafted' (*War* 178).

Perhaps the final toxic substance the soldier brings home is the sense that, no matter how bad it is and how much it deforms the self, war is also a remarkable, astounding, unmatched experience; the corollary is that civilian life is trivial and dull. War is an exciting cultural product worth spending billions on each year, despite the cost to world-wide human education and welfare, let alone the global environment. Swofford responds acidly to the idea of righteous kills: 'Unfortunately, many of the men who live through the war don't understand why they were spared. They think they are still alive in order to return home and make money and fuck their wife and get drunk and wave the flag. These men spread what they call good news, the good news about war and warriors' (Swofford 254). Swofford's anger attacks the equation of success with all forms of consumption – of material goods, physical trophies, patriotism. Then the definition of winning will be passed on to the product of the fucking. People like Grossman spread the good news, the gospel, about war and warriors with the fervent rhetoric of an orating preacher:

> Were we to go but a single generation without such men and women [warriors], we should surely be both damned and doomed. You see, we could have *all* the doctors go away for a generation and it could get real ugly, but life would keep on going. All the engineers could to away for a generation – there's lots of things that'd break down, but society could keep on going. We could have all the teachers go away for a generation and we'd have to play catch up with the next generation, but life

would keep on going. But if the *warriors* went away, if the men and women who are willing to face evil, and to march toward the sound of the guns went away, then in the span of that single generation, we would surely be both damned and doomed. (1:1)

Grossman assembles a militarist philosophy that owes debts to Plato, Hobbes, and Skinner. From Plato he borrows without any irony the concept of the Republic's philosopher kings: they are people who know justice and unworriedly mete it out. From Hobbes he selects the vision of the world as a dangerous place that cannot withstand a generation of return to the state of Nature, the war of all against all. From Skinner he extracts the lessons of operant conditioning, of teaching his subjects not only to subordinate their emotions to reason (Plato again) but to desensitize the sympathetic nervous system through endless training. Called in to analyse the shootings at high schools in Paducah, Kentucky, Springfield, Oregon, and Littleton, Colorado, he testified to the House of Representatives and the Senate that first-person shooter video games have no purpose other than to train killers (not warriors). The killer commits atrocities and attacks sheep, the warrior protects the flock. Even if the warrior must kill children (and increasingly the world's armies are child armies [Boothby and Knudsen]), Grossman's training will stabilize the conscience ('It can be a little bit hard,' he says about killing children, but 'it's not something overwhelming' [2:1]). The righteous kill, based on a chivalric code Grossman draws from the myth of the paladin, a knight, not vigilante, protecting the defenceless, returns his audience to a pre-modern world based on military authority: that only one Arthurian knight ever saw the Grail is lost on Grossman. Setting aside the idea that cultures might operate differently, that value and worth can be constructed in other ways, Grossman puts deadly force first, before medicine, engineering (building, invention), teaching (and so, learning). One thing tops Grossman's hierarchy of needs: killing.

Since so many veterans record a love-hate relationship with their war experiences, will their children really want to go to war? Grossman wonders rhetorically: 'Where do we get such men and women? We *build* them. We build them and nurture them one step at a time' (1:1). He means 'build' through training. The self-selected sheepdogs will be brought together, desensitized to the (proper) use of violence, instructed in the moral code as defined in the mission's Standing Rules of Engagement, and released into the pasture. But Grossman's model requires time. His peers in the military-medical establishment have long been working towards fast chemical programming. In 1987 Major Richard Gabriel warned that the 'main directions of military psychiatry are pointing to a chemical solution to the problem, the development of drugs that will prevent the onset of battle-shock symptoms,' with the potential of a 'chemically changed soldier whose mind has been made over in the image of a true sociopathic personality' (7). Chemical solutions have been military-medical goals for more than sixty years: in 1940 the untried new drug sodium amytal was given to shell-shocked soldiers in order to induce coma. On waking (some didn't), the subjects seemed refreshed and unafraid (the effects were short-lived, and sodium amytal was soon retired from combat). Even older drugs like dextroamphetamine (or Dexedrine, now called 'go-pills' by the U.S. Air Force) are still present in battlespace, as a much-publicized 2002 friendly-fire incident in Afghanistan proved. The pilots who unloaded a bomb on unlucky Canadian troops claimed that Dexedrine distorted their ability to think clearly. The real story is that Dexedrine (a U.S. Drug Enforcement Agency schedule 2 controlled substance; schedule 1 substances include heroin, morphine, LSD), banned from the cockpit by the Air Force in 1992, was officially restored to pilots sometime later in the 1990s. Dexedrine is merely one part of DARPA's solution for a twenty-four-hour war-fighter; two programs are at work on the problem: one is Continuous Assisted Performance (CAPS), the other is Preventing Sleep Deprivation (PSD). Both seek to reduce pilots' (and ground troops') need for sleep through manipulation of circadian rhythms and the use of new chemicals like Ampakines (drugs that enhance alertness, short-term memory, and decision making). Reassurance by the Air Force that 'go-pill' use is carefully monitored doesn't negate the fact that Dexedrine represents a systemic desire for perpetually wakeful soldiers. The biggest problem the DARPA's Defense Sciences Office faces is not the chemicals or hardware but humans: 'As combat systems become more and more sophisticated and reliable, the major limiting factor for operational dominance in a conflict is the warfighter' (DARPA 'CAPS'). The telltale word here is 'reliable': no matter how complex the system of systems is, it's *still* more trustworthy than the bag of chemicals that is a human (and that DARPA reconstructs as a war-fighter). Gabriel has already warned that 'once the chemical genie is out of the bottle, the full range of human mental and physical actions become targets for chemical control' (151).

The military has put its faith in scientific rationalism, planning, engineering, information technology, biochemistry, and biomedicine. It has in every way so far possible applied physics, chemistry, math, engineering, and the life sciences to war in order to get more product (victory, territory, resources) for less cost (dead people, lost machinery – wastage). No participant understood this more clearly than Dow Chemical, besieged by sick veterans. One Dow spokesperson said before the out-of-court Agent Orange settlement: 'We think that 2,4,5-T [a key ingredient in Agent Orange] is a very important symbol. If we were to lose on this issue, it would mean that American public customs would be beaten back a couple of hundred years to an era of witch hunting. Only this time the witches would be chemicals, not people, and that's the importance of this issue' (quoted in Wilcox *Waiting* 144). Dow did 'lose' in that it settled, but Americans have hardly lost faith in science. There were two parallel victories. 2,4,5-T represented the U.S. technological victory over previously impenetrable triple-canopy Vietnamese jungles; depleted uranium, now thoroughly accepted by the world-wide arms community, represents the triumph of business over fundamental health issues. DU is proof that the Agent Orange lawsuit representing 2.4 million people and paying out a miniscule $180 million USD (the plaintiffs sought an apology and from $4–40 billion USD in reparation) was a sideshow. DU, the nuclear garbage of the United States, is more powerful than some of the world's most formidable weapons (Soviet T-72 Warsaw Pact era tanks). The American weed-killers 2,4,5-T and 2,4,-D pack more punch over the long haul than even the National Liberation Front's demoralizing victory in Vietnam, where the chemicals have gone on killing Vietnamese civilians and their children. Pyridostigmine bromide may cause a few rashes, but it isn't anything science won't fix. And if science can't fix it and there's more wastage – it won't be anybody anyone knows.

The new war-fighter is a science fantasy dreamed by insomniac DARPA. The agency seeks 'Metabolic Dominance,' the name of one of its recent programs, over the human machine's central-processing unit. Like the sleepless pilot, the ideal ground soldier is one who needs little rest and, ideally, no food. Troops move better the less they carry, so the ideal soldier will need no pack, food, or water. The war-fighter has already been reclassified by the doctors at Walter Reed Army Medical Center as a 'tactical athlete' (O'Driscoll): Why wouldn't that athlete's coach want the ultimate power-drink for the team? DARPA's public tender for the 'Metabolic Dominance' program has now entered the second phase, which means that scientists are currently working towards 'the ultimate goal [which] is to enable superior physical and physiological performance by controlling energy metabolism on demand. An example is continuous peak physical performance and cognitive function for 3 to 5 days, 24 hours per day, without the need for calories' (DARPA 'Metabolic'). The search is on for a skin-patch or slow-release pill that will, over a number of days, continuously recalibrate the metabolism and enable high-level sleepless performance (absent the irritating side-effects of paranoid delusions that accompany most amphetamine use). If soldiers could have their systems over-cranked this way, three to five days of non-stop fighting would, in DARPA dreamings, end the war. The body machine will need a full overhaul, particularly in order to control 'fluctuations in core body temperature [that] limit warfighter performance' (cooler and steadier core temperatures make for less caloric uptake and greater physical endurance). Nobody knows what the long-term cost of 'metabolic dominance' to the body, cognition, and the immune system will be. But it is certain that if offered a substance that promises sleeplessness, strength, and reduction of combat time, many soldiers will gladly take it. They are, after all, units in a machine that accepts war chemicals as an 'important symbol' – or reality.

Of the 11,069 American soldiers wounded to date (over 120 with limbs lost) in the U.S. wars on Afghanistan and Iraq, 'up to 40% of all of those injured will be able to return to active duty,' says the chief of the amputation wing at Walter Reed (O'Driscoll). O'Driscoll adds that 'in today's military, amputation doesn't automatically mean "medical retirement,"' a fact Joe Haldeman foresaw in his 1975 *The Forever War*. Protagonist William Mandella, having lost a leg and expecting his war to be over, discovers that medical engineering has learned to regenerate whole limbs and nerves: back to the war. Returning to combat, which would have been unthinkable for most amputees even ten years ago, is now possible. A Green Beret sergeant whose right hand was blown off in Afghanistan can, with a new prosthetic, field-strip and reassemble a weapon in ninety seconds: nothing will keep him and his high-tech limb from Special Forces (O'Driscoll). New computerized prosthetic legs calibrate the soldier's foot-strike and balance the body on rough surfaces. What if a soldier wishes to leave the field with such a sophisticated, expensive limb? Will the soldier be asked to trade in the limb for a more standard steel prosthetic? What if the soldier thought that shooting off a toe, foot, or trigger finger would cause an army discharge, only to find that it is not so? What if the soldier can't have the new limb unless he works (for the army) to pay it off?[51] Medical engineering is not-so-slowly closing up the gaps for the resisters, those who haven't decided, in Grossman's dictum, whether they are warriors or victims.

One Vietnam veteran, son of a Second World War veteran, concludes that 'everybody has an obligation to their country – I agree with my dad on that. However, not my kids. I hate to say that. I know it sounds like a double standard. But I've seen what they do' (Takiff 384). What they do. What the whole system, now mobilized for 24/7/365 war, does when it overcomes the body, and then the self. The skin, useless against the military's remarkably invasive pliers, cannot prevent mechanical, chemical, psychological wires from entering the body, the brain. The re-created flesh must be trained and primed to survive battlespace, to match its tempo, to cycle into a war metabolism. Body armour will be one of the first contacts the war will insist upon, and then, in order to establish larger safe zones, it will turn to rolling armour. Extended by technology, the personal body expands into new territory: the body will need constant upgrading in order for it to act as a testbed if it is to survive the gun-armour spiral, the missile-tank-helicopter roundabout. The body twists into the letter 'X,' the blank signifier used by the military to indicate that which has not been fully formed, that which is always about to become: the trial version of the new killer application. The X-human will be inscribed so that it is survivable through those first three to five days of perpetual sleepless combat. During those hours, the air, ground, water, the whole environment will be patrolled by Von Neumann machines replicating themselves for vigilant autonomous duty. Because of nuclear, biological, chemical threats, tissue will need to remain in hardened containers that support constant over-pressure, where air pushes out toxic substances, including the ones the military itself brings into battlespace. Open-flying dune-buggies have vanished in favour of hovering tanks, gun-platforms and ductile craft, prowling the lower air-deck. Around them and through them all flashes the currency of C4ISR, interoperable joint war-fighting, SIGINT, MASINT, and HUMINT, all connected on the worldwide war web, which has destroyed all information stovepipes, pillars of discrete and now useless data.

Far behind lies the empty skin-bag wastage. All around it extends war.

Notes

1 Wondering if he would be called on to decide about using nuclear force, President Lyndon Johnson asked U.S. Army Chief of Staff General Earle Wheeler to look into the pros and cons of hitting Vietnam with atomic weapons. Wheeler passed the job to General William Westmoreland, then overall commander of troops fighting in Vietnam, who formed a secret group to study the issue. He was ultimately told 'to desist' by Washington lest the press hear about the group. Westmoreland later wrote irritatedly: 'I felt at the time and even more so now that to fail to consider this alternative [nuclear weapons] was a mistake' (445). Use of nuclear weapons was urged by 1954 chair of the Joint Chiefs of Staff Arthur Radford (Bowman 12) and later by Kennedy's deputy special assistant Walt Rostow and political opponent Barry Goldwater (Baker 374).

2 Despite its attempt to use gender-neutral language, in the area of robotic vehicles the defence establishment refers to all air, land, or sea craft without operators (female or male) as 'unmanned.'

3 Vietnam veteran Joe Haldeman captures the soldiers' mixture of rage, relief, horror, and detachment at seeing a napalm strike; the men watch from a distance 'before the roasting smell' hits them: '*burn baby burn* – we'd say' ('air support'). Later, many Vietnam veterans were unable to tolerate the smell of cooked meat.

4 Becker's comment is problematic because it uses the ambiguous phrase 'total war.' For some scholars, total war is defined as a global nuclear conflict that would end the planet's ability to support life. For others, total

war is defined in opposition to limited war, which is fought for a specific goal, is restricted to soldiers, and does not extend to nearby civilian populations. I understand a total war to be one 'in which the whole population and all the resources of the combatants are committed to complete victory and thus become legitimate military targets,' including the 'assymetrical total war in Vietnam and a number of other post-imperial conflicts,' where the 'urge to win means there is no natural resting point in what we now call "escalation"' (Bicheno 915–16). Total war can be marked by scorched-earth policies, where it is better to destroy the environment than to let the enemy, or civilians, use it. It would be tempting to label a number of limited wars as total, and Bicheno warns against just such a tendency. The reader interested in further discussion of this problematic term might explore Stig Förster and Jörg Nagler's edited collection *On the Road to Total War* (New York: Cambridge University Press, 2002).

5 Hoping to escape fighting in the undeclared war on Iraq, two American soldiers fled to Canada in December 2003, where they applied for refugee status. The Canadian government has considered sending the two men back, but they now have a lawyer and the legal case is under way. One army spokesperson said of the ex-soldiers' resistance: 'It goes against the Army's values' (Williams).

6 An American soldier in Iraq writes that 'it's hard listening to my platoon sergeant saying, "If you decide you want to kill a civilian that looks threatening, shoot him. I'd rather fill out paperwork than get one of my soldiers killed by some raghead"' (Moore). Shooting and staying alive is better than holding one's fire and going home in an aluminum case.

7 'Smash and run' is eerily anticipatory of the typical U.S. Army 'search and destroy' (later 'sweep and clear') missions in Vietnam.

8 Nanotechnology refers to engineering that goes on at an atomic level (a nanometre is a billionth of a metre). In the near future, nanomachines will be applied to most scientific endeavours, including non-invasive surgery, treatment of cancer and heart disease, information technology, and military technology.

9 Van Creveld traces the emptying of the battlefield from the Greek phalanx (one soldier per metre [1:1]) to the eighteenth century, where the soldier-to-ground ratio 'had decreased to perhaps 1:10,' through the American Civil war (1:25), the First World War (1:250), and the Second World War (1:750) (*Technology* 173).

10 The term 'Fuel-Air Explosive' is unfamiliar to most of us, but we may recall the so-called 'bunker buster' (a 2,000 kg bomb) developed at the end of the first Gulf War for an attack on the 'hardened' bunker system that Saddam Hussein's Republican Guard was said to be dug into. Bunker busters were used extensively in the American carpet-bombing of Afghanistan in 2001–2.

11 Secret weapons are often designated by numbers before they're given names. Tanks were originally tested in England, and, if anyone asked what the strange canvas-draped objects were, they were told that they were new types of water carriers or cisterns: 'tanks.' Before that, there wasn't a generic word for them in any language. Robert Graves records his early encounters with poison-gas canisters that the British forces referred to as 'the device,' in part for secrecy, in part to acknowledge the unspeakable quality of gas warfare. UAVs were, for a long time in the Cold War, called SAPs (Special Air Projects) or referred to by their aircraft designation letters BQM, a code for uninhabited air vehicles. All this secrecy flew in the face of the fact that the British public were subject to the very public and successful displays of V-1 and V-2 robot aircraft.

12 The idea of weapons from the future is at the heart of Joe Haldeman's *The Forever War* (1975). Because Haldeman's soldiers experience time dilation as they travel through space, they are never sure whether they will arrive at a distant battlefield technologically ahead of, or behind, the enemy. Time dilation operates not only as a metaphor for war's wrenching actions on human senses but also as ironic commentary on the American war in Vietnam, where Haldeman witnessed the destruction of a technologically superior force by the fiercely intransigent, agrarian Vietnamese. General Curtis LeMay wished to 'bomb them [the Vietnamese] back to the stone age,' although presumably he considered they were not far from it.

13 One American tanker explained irritably: 'Our commanders have decided on a new tactic. If the Germans send a Tiger tank we will send out eight Shermans to meet it and we expect to lose seven of them.' http://www.panzer-vi.fsnet.co.uk/tigerphobia.html (accessed 2 July 2003).

14 Modifying weapons is considered to be illegal by the Army and by extension the state because the alterations are made by soldiers, not weapons'

designers or knowledgeable technicians: the modifications can and often do kill their creators and fellow soldiers. Adding armour and weight to a tank, while it seems to make sense, can burn out its motor or break an axle, a problem if the tank commander has come to rely on the machine's manoeuvrability to carry it out of enemy gun range. In Vietnam, soldiers facing the impossible task of keeping the finicky M-16 rifle clean resorted to using gasoline to wipe down the gun's parts. The practice was illegal because gas fumes could explode when the rifle fired; on the other hand, the guns had to be cleaned and the Army knew that the M-16 was a dangerous lemon, so the practice was officially forbidden but unofficially permitted. The now-famous 2004 case of the soldier in Iraq referring to 'hillbilly armour' in an exchange with Donald Rumsfeld is one of the most recent instances of battlefield modification. Bad design is a killer.

15 Depleted uranium has been given away, according to its critics, at almost no charge to munitions makers. It was used not only for tank sabots in the first Gulf War but also for bullets in AH-64 Apache helicopters and A-10 Warthog machine-guns. The British Royal Society replied sharply in April 2003 to the Pentagon's claim that there was no need to clean up the some 2,000 tons of DU expended by American forces during the second Gulf War (2003) (Brown 'Scientists Urge Shell Clear-up').

16 Stalin, finished with his disastrous pre-war purge of Russia's experienced generals, finally established Tankograd (the industrial city of Chelyabinsk), whose sole job was to make tanks. Setting apart the other war machinery produced at Tankograd, some eighteen thousand tanks rolled off the line there during the war. http://www.globalsecurity.org/military/world/russia/chtz.htm. Accessed 16 May 2003.

17 Early British tanks were designated male or female depending on the armament they carried.

18 The dreadful image of humans being broiled in slow- or pressure cookers is worth keeping in mind when we consider the infamous 'Zippo raids' that Americans routinely performed in Vietnam, where Zippo lighters were used to start fires that eradicated Vietnamese villages. Again, a familiar domestic object of comfort and relaxation, the Zippo lighter, becomes not only a weapon of war but a tool in what some now consider to be atrocities.

19 Worrying about the cost of industry, Thoreau asked his *Walden* readers almost a hundred years before: 'Did you ever think what those sleepers

are that underlie the railroad? Each one is a man, an Irishman, or a Yankee man. The rails are laid on them, and they are covered with sand, and the cars run smoothly over them. They are sound sleepers, I assure you. And every few years a new lot is laid down and run over; so that, if some have the pleasure of riding on a rail, others have the misfortune to be ridden upon' (174). Thoreau plays on sleep: sleepers are railroad ties, as well as humans who have fallen asleep in their lives. The reality is also that thousands of labourers died to make railroads, and their bodies become the metaphorical road on which the machine runs. Industry is hard on humans.

20 The human 'bump' is now a trope. One of the more thoughtful lieutenants in Alex Vernon's book orders, in frustration over trying to straddle a bunker, that his tank '"Hit the bunker ... Crush it." We hardly noticed the bump' (219).

21 MBDA is a multinational corporation comprised of French (Matra), British (British Aerospace [BAe] Dynamics), and Italian (Alenia Marconi Systems) missile builders (other partners include Sweden [Saab], Germany, and Spain).

22 Rumsfeld's belief in the RMA was heard most palpably by North Americans when, in the winter of 2003, the phrase 'shock and awe' suddenly appeared in every news source. *Shock and Awe* was written in 1996 by early RMA architects Harlan K. Ullman and James P. Wade. Iraqis alive in 1991 were already familiar with the term.

23 The corollary this comparison suggests is that, just as the horse cavalry had to give way to fully mechanized armies, so will heavy forces be compelled to step aside for light strike forces. However, the comparison makes no such argument. It's about the affection the soldiers have for their various kinds of equipment (horses, heavy armour), not the similarities between those pieces of equipment.

24 The 2003 Iraq War, declared over by President George W. Bush on 1 May 2003, could be used as an examplar of RMA warfighting: it was fast, brutal, high-tech, interoperable, and apparently decisive. However, RMA theory and desert reality didn't quite match. By September 2004, a defensive President Bush called the military victory a 'catastrophic success,' a phrase one part Orwell, one part RMA rhetoric. As in Vietnam, killing large numbers of the enemy didn't make for victory, and high technology didn't prevent U.S. soldiers from having to stay on the

ground where they provided steady targets for forces fighting a uncon-
ventional resistance. General Eric Shinseki, Army chief of staff from
1999 to 2003, enraged Rumsfeld because he warned the Senate that
Rumsfeld's war would require well over 200,000 troops in order to hold
Iraq. Late in 2004, Paul Bremer, once administrator in occupied Iraq,
agreed with Shinseki's assessment, but he retracted his statement a day
later. The phrase 'catastrophic success' continues to haunt anyone looking at the ever-rising civilian and soldier casualty figures.

25 The phrase 'the fog of war' is usually attributed to Clausewitz, but he
never said it. The fact that some military theorists find it too vague to be
useful hasn't prevented it from appearing in part or whole as a title for
books and films and as an accepted part of strategic discussion. It is usually understood to mean the chaos that makes the battlefield into a
murky, impenetrable zone.

26 Few works of art take up these overlapping myths of military, mechanical craft, and frontier persistence as brilliantly as Buster Keaton's 1927
The General, in which a quiet train-loving engineer during the American
Civil War fights what seems to be the entire Union army in order to save
the South, win his girl's heart, and, most of all, protect his beloved train
engine *The General* (which he must sacrifice to win the girl).

27 The war over the particular direction in the Revolution in Military
Affairs is perhaps best laid out by Elinor Sloan in her book *The Revolu-
tion in Military Affairs* (Montreal and Kingston: McGill-Queen's Univer-
sity Press 2002) and in Chris Hables Gray's 'The Perpetual Revolution in
Military Affairs,' in Robert Latham, ed., *Information Technology and
International Security* (New York: New Press 2003).

28 The surreal quality of a popular culture that takes its cues from war
seems endless. In the 1970s ballpoint pens called 'Doodlebugs' were
manufactured and sold in North America without any apparent irony. A
further bizarre turn of events has seen 3M trademark 'Doodlebug' for a
line of floor cleaners. Bombs appear to make good product names, from a
video-store chain named after the Second World War bombs designed to
'bust' whole city blocks to scrubbers that presumably kill dirt.

29 Currently, the United States is investing in what it calls brilliant weapons.
The BAT (Brilliant Anti-Tank weapon) is fired from a 155 mm howitzer.
Once aloft, as with the SADARM (Sense and Destroy Armor), the BAT
pops out a balute (combination of balloon and parachute) and begins a

slow oscillating descent as it seeks targets (usually tanks). Another such weapon is the Quicklook, a shell that, once airborne, sends back imagery to receivers and then crashes. While these weapons borrow a great deal from UAVs, they remain part of the munitions world.

30 For retrieval, the relatively new SeaScan and ScanEagle fly right at a taut wire that snags an upraised wingtip. It's hard not to wince as the UAV slams into the line and spins around it, dangling painfully, but the method is similar to the tailhook-and-cable system used to land jets on aircraft carrier decks.

31 Interlaced screens, like most current television sets, tend to flicker. The flicker comes from the way information is 'painted' on the screen by the electron gun at the back of the tube. In the first pass, the gun paints half the screen (say, all odd lines), and on the next pass it paints the rest (the even lines). The electron gun's scanning occurs sixty times a second. Because they are 'on' all the time, liquid crystal displays or plasma screens are not interlaced and so don't flicker, providing sharper images for human eyes.

32 By the later days of the Vietnam War, Air Force ardour had cooled a good deal. Vietnamese prisons in the North were full of American pilots waiting out an apparently endless war. Dale Schwanhausser, central to the Ryan mission in Vietnam, recounts: 'From his offset position, the U-2 pilot saw the Lightning Bug drone shot down. Back at the Officer's Club later at Bien Hoa, he told me how he watched the "telephone pole" [surface-to-air missile] consume the drone. The rivalry stopped when he said, "From now on, you guys can have that mission"' (Wagner 99).

33 A human in the net is an uneven lump in the web; netcentric warfare (as conceived by Vice Admiral [Ret.] Arthur Cebrowski) specifies that 'a node can be an airplane, a general, an Army private, a tank, or a UCAV' (USAF Scientific Advisory *New World Vistas*).

34 Bandwidth capacity refers to the amount of information that can be sent at given speeds. Talking about bandwidth is like talking about a water or gas pipe: the larger the pipe, the greater the capacity for material delivery (whether the material is a fluid or information sent as pulses of data). On the ground and inside computers, dielectric mirrors promise to surpass even fibre optics for speed and ease of use.

35 The use of the words 'family' and 'class' suggests a Linnaean taxonomy (incomplete) for uninhabited vehicles.

36 When the American Army engages in war games, the two forces are divided into Blue (friendly) and Red (presumably a holdover from the Cold War) forces. The United States forces are permanently encoded Blue, so blue-on-blue attacks are what the press calls 'friendly fire' incidents, not unlike the one involving some Canadian soldiers in Afghanistan, who were bombed by U.S. pilots in the summer of 2002. That was a blue-on-blue situation.

37 Recall that a terabyte is a thousand gigabytes. By 2010, UAVs will be flying with more than enough processor and storage power to handle the 'map of the world' thanks in part to dielectric mirrors and new messaging systems currently being worked on by Advanced Technologies and Boeing (Fulghum 'Dumb Blobs' 47).

38 United States military policy once more reveals itself. Prepared to use depleted uranium on foreign soil and leave radioactive dust for the inhabitants to breathe is as much nuclear contamination as the willingness to fly a nuclear power-plant 'over foreign countries.'

39 A half-life is the time it takes for a given material to lose half its radioactivity. Hafnium's half-life is thirty-one years, so, at the end of that period, half of its radioactivity will be gone, in another thirty-one years, half of that half; and so on. Some radioactive substances take more half-lives (essentially, generations) than others to become benign. Caesium-137 and Strontium-90 (both human-made radioactive substances) have half-lives of thirty and twenty-eight years, respectively. Both are virulent carcinogens, with Strontium-90, a bone-seeking agent, responsible for a multitude of bone cancers.

40 Filmmaker Michael Moore's first e-mail to his website subscribers after George W. Bush's re-election was a minimalist statement of horror, beginning with the line 'my first thoughts after the election' and followed by a full list of the then over 1,100 U.S. war dead from the second Gulf War (Moore).

41 In Vietnam, the slang term 'pogue' can be used sympathetically towards grunts – a pogue is what the British in the First World War would have called the 'PBI' (Poor Bloody Infantry). The second meaning points towards 'punk' in the sense of a young, inexperienced (possibly homosexual) fool. 'Lifer,' especially during the Vietnam years, was a virulently nasty term for soldiers who were career army and in 'for life'

(usually twenty-five years). 'Pogue lifer' represents the brutally creative language that is a hallmark of enlisted life.

42 According to Dave Grossman, there *is* an ambiguity about the sixth commandment. Grossman argues that the King James Bible mistranslates it as 'Thou shalt not kill' when in fact the proper translation is 'Thou shalt not murder.' Into the eye of this needle Grossman stuffs the camel of religious justification for battlefield 'righteous kills.' The Jerusalem Bible, a plain-language Bible, translates the commandment as 'You shall not kill' (Exodus 20:13).

43 No doubt Fussell angered many when he wrote in the same autobiography, 'Thank God the troops, most of them, didn't know how bad we were. It's hard enough to be asked to die in the midst of heroes, but to die in the midst of stumblebums led by fools – intolerable. And I include myself in this indictment' (*Battle* 173).

44 In a number of war texts (books and films), returning soldiers are healed when they come in contact with Native Americans; Western medicine is shown to fail where Native cleansing rituals succeed. See, among other works, two books – Leslie Marmon Silko's *Ceremony* (1986), and Philip Caputo's *Indian Country* (1988) – and two films – Sean Penn's *The Indian Runner* (1991), and John Woo's *Windtalkers* (2002).

45 Rudyard Kipling wrote the famous poem 'Common Form' about his son's death at the Battle of Loos (1915): 'If any question why we died,/ Tell them, because our fathers lied' (published 1919).

46 Various anti-war forces working to stop the invasions of Afghanistan in 2002 and Iraq in 2003, had to step carefully around the question of soldiers' complicity. Before the Iraq ground war in early 2003 there was dissent about the cause even from the right wing. Once the war began, doubters largely fell silent because speaking against the war and the troops were seen to be the same thing. One of the lessons the country learned during Vietnam was not to blame its soldiers for a bad war. However, the equation of the troops with the war worked nicely to choke off public resistance to the war: attacking the war meant assailing the largely working-class soldiers sent to fight it. Even as the war and occupation quagmire lose favour and staunch Republican families turn against Bush White House war policies, peace advocates who want to be heard must repeat that it is the war, not the soldiers, they are unhappy about. Prison

scandals at Abu Ghraib and Guantanamo made such a distinction that much harder to sustain.

47 'Charlies': American slang for North Vietnamese enemies known as Viet Cong – VC – who on the military radio called in as 'Victor Charles,' hence 'Charlie'; 'Sammies': Somalians, sometimes known by the American troops as 'Skinnies' – also the name for one of the worthless alien races in Robert Heinlein's *Starship Troopers* (1959).

48 Before the fall 2004 presidential election, the GAO acronym was altered; what had been the General Accounting Office became the Government Accountability Office.

49 The Government Accountability Office has since uttered a scathing report of the Department of Defense's simulation of the number of veterans exposed to toxic substances in the Gulf (Rhodes).

50 Dr Alvin Young, an experienced army chemist whose professional career was based on defending phenoxy herbicides, said about the limited understanding of these chemicals: 'When I first got into the herbicide business [1960] ... we weren't concerned about toxicity ... My heavens, we didn't consider the phenoxy herbicides toxic. We sprayed each other ... as a game, and we would go to our supervisor and say, are these things toxic and the answer always was, oh, no, no, no. Herbicides are not toxic' (quoted in Severo and Milford 379). Young has since established an archive at the National Agriculture Library dedicated to the problems of phenoxy herbicides.

51 In case these questions seem absurd, it's sobering to note that the Pentagon, through President Bush, has issued a series of 'stop-loss' orders that prevent soldiers who have completed any number of tours (or years) in Iraq or Afghanistan from leaving the army (Lumpkin). The result is that the army has exceeded its legal personnel limit set by Congress (the forces aren't supposed to rise above 485,000 people – stop-loss orders have put the numbers over half a million). The National Guard fell short of its 2004 recruiting quota by some 5,000 people. In 2003 the Democrats brought a conscription bill to the House in order to force the government to demonstrate that it would not initiate a draft. The bill languished for eighteen months until October 2004, when, with rumours on college campuses growing that there would be a draft, the Republicans suddenly put the bill on the docket and voted overwhelmingly against it (the vote was 402–2). As one story noted: 'Laurie Rivlin

Heller, of Mothers United to Stop the Draft, called yesterday's House vote "a public relations stunt." The question isn't whether there will be a draft next year, she said, but in two or even five years' (Babington and Oldenburg). Her guess is supported by Vietnam veteran David Hackworth, who concludes, 'You don't have to be a Ph.D. in military personnel to conclude we need more boots on the ground' ('Uncle Sam'). The issue of stop-loss orders, called a back-door draft by some, has refused to subside, and it has been further heightened by the recall of soldiers and reservists who completed their military duty. At least one veteran of the first Gulf War, David Miyasato, who had been recalled to service after being honourably discharged over ten years ago, decided that enough was enough and took the Army to court. He won the case, but both he and his lawyer voiced concerns over what are thought to be 4,000 other ex-soldiers or reservists in his position (Goodman).

Three years after the start of the ground war in Iraq, the draft is now inevitable should the White House intend to maintain current troop strength. The news that the Selective Service System could go into operation on 14 June 2005 has done nothing to alleviate the fears of potential draftees and their parents, children, husbands, and wives.

Glossary

Access 5 Program: Joint government-defence industry program to give UAVs routine access to North American airspace by 2008.

ADWs (Active Denial Weapons): Tunable microwave weapons intended for use on human beings. New research is focused on creating intense short-term pain in the enemy, or crowd, using Pulsed Energy Projectiles, or PEPs. *See also* HPM.

AirLand Battle: Quick, unpredictable way of fighting according to RMA doctrine.

ALSC (Accelerated Life Simulation Computer): Full-body virtual reality immersion tank used to train soldiers. Invented by science fiction author Joe Haldeman in 1974, now close to being fulfilled.

APC: Armoured Personnel Carrier (sometimes 'track').

APFSDS (Armour Piercing, Fin Stabilized, Discarding Sabot): Particularly lethal tank round at the heart of which is a long-rod depleted uranium, sometimes tungsten, penetrator.

Asymmetric War. *See 4GW.*

AWACS: Airborne Warning and Control System. Relatively high-flying aircraft whose chief task is to handle air traffic and monitor enemy planes in battlespace.

BAMS (Broad Area Maritime Surveillance): Part of the U.S. Coast Guard's program to keep twenty-four-hour watch over the Atlantic and Pacific coastlines, as well as the Rio Grande (much surveillance will be carried out by UAVs).

BATs (Brilliant Anti-Tank missiles): Autonomous missiles that, even in bad weather, can see, hear, and choose a target, then guide themselves to that mark.

Battlespace: The entire area in which battles are fought, from space to ground, over and under water, and across time.

BMIs (Brain Machine Interfaces): Brain impulses are detected by a computer attached to, or implanted in, the human; the machine interprets the signals and relays commands to a robotic aircraft, land vehicle, arm, or weapon.

Brilliant weapons: Autonomous munitions that can be programmed for various tasks.

CAPS (Continuous Assisted Performance): Program designed to allow soldiers to fight with little or no food or water for two to three days at a time.

CCTT (Close Combat Tactical Trainer): Networked virtual training system that allows a squad of soldiers to play war-simulation games as a unit. The CCTT may be a network of desktop computers or sophisticated wrap-around simulators.

C4ISR (C^4I^2): Command, Control, Communication, Computing, Intelligence, Surveillance, Reconnaissance.

Chobham Armour: Armour said to be able to defeat explosive tipped weapons, and perhaps even sabot rounds (unknown).

COTS (Commercial Off the Shelf): Commercial machine and computer components used to build new weapons: commercial parts are cheap, standardized, and, typically, plentiful (unlike special machine pieces made by hand one at a time). May also refer to software.

Cyberwar: War for domination of the intelligence resources: hacking, cracking, attacking the enemy's Internet and computing resources, using software in order to destroy or disable those resources.

Cyborg: Cybernetic organism, a fusion of human and machine where human neural impulses govern mechanical prosthetic devices: the result is greater than the sum of the parts.

Cyclic: Main control handle in a helicopter.

DARPA: Defense Advanced Research Projects Agency. United States agency responsible for most new war technology.

Directed Energy Weapons (DEWS): Tunable weapons like micro-waves that can be directed at whole populations in order to induce them to leave an area (colloquially: 'dial-a-hurt'). *See also* ADWs, HPM.

DU (Depleted Uranium): Nuclear reactor junk by-product that is extremely heavy and dense, used for tank sabots (APFSDS) and embedded in tanks hulls. Distributed free to U.S. arms manufac-turers.

ELINT (Electronic Intelligence): Electronic signals picked up from foreign or enemy transmissions and used to determine enemy activity.

EMP (Electromagnetic Pulse): Depending on its size, an EMP can shut down or destroy all electronic equipment in a given radius. Nuclear weapons generate powerful EMPs.

ERA (Explosive Reactive Armour): Tank armour that reacts to an attacking shell with a countering explosion that offsets the attack.

ERYX: Relatively short-range, portable anti-tank missile carried and launched by one or two operators.

FAE (Fuel Air Explosives), also Thermobaric Weapons: Explosives that put up a cloud of combustible fluid or powder that forms an above-ground umbrella which is then ignited by a second explosion on the ground. The result is a vacuum near and below the blast.

FCS (Future Combat System): Theoretical designs for what the armour, rotorcraft, and aircraft of tomorrow might look like.

Force XXI: U.S. Army document that outlines the direction the forces will take in the next ten years.

4GW: Fourth Generation Warfare, sometimes *Asymmetric Warfare*. David and Goliath struggles between rebels or 'terrorists' and large states or superpowers.

GBU (Guided Bomb Unit): Laser guided 'smart' munitions that include the GBU-28 2,270 kg 'bunker buster,' which drives thirty metres into the earth before exploding. The GBUs are all laser-guided, which distinguishes them from unguided and therefore so-called 'dumb bombs.'

GDLS (General Dynamics Land Systems): Unit of General Dynamics responsible for designing and manufacturing armour like the Abrams tank and Stryker vehicles.

GIG: Global Information Grid. The U.S. Department of Defense's renewed military Internet that operates almost exclusively over fibre-optic cables, providing high-resolution real-time data transmission.

Glacis: Tank's front slope as well as slope in front of a typical fortress.

GWI, sometimes GWS (Gulf War Illness, Syndrome): Constellation of debilitating ailments (fatigue, memory loss, chronic headache, joint pain, dizziness) that has crippled over 200,000 U.S. veterans of the first Gulf War.

HALE UAV (High Altitude, Long Endurance): Highest and longest flying class of UAVs (including Global Hawk and Helios).

HDTV (High Definition Television): Digital television displays that are roughly twice as sharp as regular television sets, free of the usual screen flicker.

HMD (Helmet Mounted Display): Generic term for any display system mounted on the helmet that allows the soldier to send and receive information. *See* IHAS.

HPM (High-Powered Microwave): Tunable microwave weapons designed to bring down enemy communications and computing systems in 'chip-frying' missions. *See* ADW, Directed Energy Weapons.

HSM (High Speed, Manoeuvrable): Class of UAVs typically designed for combat.

HUD (Head Up Display): Sight that superimposes targeting and navigation information on a clear screen in front of the pilot, who thus need not look down at the instrument panel.

Huey: Bell UH-E1 helicopter, familiarly known as a 'Huey.'

HUMINT: Human Intelligence: People on the ground who provide forces with usable data. *See* MASINT, SIGINT.

IADS (Integrated Air Defense System): The electronic eyes and ears of enemy forces.

IEDs (Improvised Explosive Devices), also Vehicle Borne IEDs (VBIEDs): Shells or plastic explosives usually buried at roadsides and rigged with switches or timers to go off in proximity to convoys or simply explode arbitrarily.

IHAS (Integrated Helmet Assembly Subsystem): Simple to complex helmet-mounted displays that show the user tactical grids, maps, and video from other soldiers or pilots. IHASs may be connected to weapons systems that track their targets by following the soldier's eye movements.

Interoperability: Sharing forces, information, and materiel as needed to achieve a military goal.

IT: Information Technology – computers, Internet, local area networks, cellphones and land lines, digital signal senders and receivers.

Jamming: Shutting down information technology – radars, radios, Internets, satellite feeds.

JATO (Jet Assisted Take Off): Also called a 'bottle,' a small rocket pack attached to a UAV so that it can take off without a runway.

Jointness: Similar to interoperability, emphasis on interservice cooperation.

Joint Vision 2020: U.S. military document explaining the concepts of joint operations and interoperability.

JSTARS: Joint Surveillance and Target Attack Radar System. Aircraft dedicated to providing ground troops with intelligence and to passing information between small units.

LAV (Light Armoured Vehicle): Usually six-wheeled, rather than tracked, vehicle ideal for speed, manoeuvre, and low-gas consumption. LAV armour cannot withstand a tank round or advanced missile (ERYX, MILAN).

LEWK (Loitering Electronic Warfare Killer): Smart munition that can loiter on station as it waits for a target.

LIC (Low Intensity Conflict): Military operations not classified as wars and expected to be relatively short and casualty-free.

Littoral warfare: Operations based in littoral environments where naval craft provide bases and staging grounds in the absence of friendly countries where troops can mass.

LOH or LOACH (Light Observation Helicopter): Small two-person rotorcraft designed primarily for observation, speed, and manoeuvrability.

LOS (Line of Sight): Limit on hand-launched rocket and guided missile: the soldier can shoot at only what can be seen. New missiles will operate Beyond Line of Sight (BLOS).

MASINT (Measurement and Signature Intelligence): Information about attackers derived from signatures that they produce (thermal, chemical) and that are detected by scanning technology. *See* SIGINT, HUMINT.

MAVS (Micro-UAVs): UAVs ranging from the size of a raisin to a 15 mm saucer.

MEMS (Microelectromechanical systems): Miniscule machines created by an etching rather than machining process. MEMS allow for extensive miniaturization and the creation of 'systems on a chip.'

MILAN: Long-range, portable anti-tank or anti-aircraft missile.

MLRS (Multiple Launch Rocket System): Mobile rocket batteries usually consisting of twelve rocket tubes, mounted on truck beds that can be elevated to give the missiles greater range.

MOUT: Military Operations in Urban Terrain. Deeply feared because so costly in human lives.

MRSI (Multiple Rounds, Simultaneous Impact): Fired in quick succession from one howitzer, a salvo that arrives simultaneously at the same target (known in the artillery during the Second World War as 'time on target').

NAMRL (Naval Aerospace Medical Research Laboratory): Unit responsible for equipping humans to perform better in the cockpit's high-tech environment.

Nanotechnology: Technology built on tiny machines measured in billionths of a metre.

NATO: North Atlantic Treaty Organization.

NBC: Nuclear, Biological, Chemical weapons.

Netcentric, Netcentrism: Organizing force structure as a network rather than top-down command; using the Internet to maintain that structure.

OFW (Objective Force Warrior): Ultimate soldier designed to operate in the field by 2010. OFW follows the Land Warrior Program and is superceded by the Future Force Warrior (FFW). OFW is a fully networked soldier equipped with the most advanced weapon, medical, stealth, and information technology.

OICW (Objective Individual Combat Weapon): Replaced by the Heckler and Koch XM8 and XM25, the OICW was a combined smart rifle and grenade launcher with computerized sight and programmable ammunition.

OOTW: Operations Other Than War – peacekeeping, humanitarian aid, relief (may also involve killing).

PB: Pyridostigmine bromide – pills given to prevent death by the nerve gas Soman.

PEO STRI (Program Executive Office Simulation, Training, and Instrumentation): Department of Defense unit charged with overseeing and developing extensive, realistic, virtual-training programs for different branches of the military.

PIAT (Projector, Infantry, Anti-Tank): Awkward spring-loaded Second World War British version of the American bazooka – designed as an anti-tank weapon. Back-breaking.

Postmodern War: War conducted according to the tenets of RMA.

PSD (Preventing Sleep Deprivation): Connected to CAPS, study aimed at reducing or eliminating the infantry's need to sleep during long periods of combat.

PTSD: Post-Traumatic Stress Disorder.

Pure War: Complex term coined by theorist Paul Virilio, similar to Postmodern War.

RCM (Reciprocating Chemical Muscles): Machine 'muscles' driven by a chemical cocktail: as the muscles work they return power to a small battery that drives the craft.

RHA (Rolled Homogeneous Armour): Until the advent of Chobham Armour, the toughest, most effective steel plate developed for tanks.

RMA: Revolution in Military Affairs.

RPG-7 (Rocket Propelled Grenade): Soviet-designed anti-tank weapon which was fielded in the early 1960s and which made the infantry once more capable of defeating tanks.

RPV (Remotely Piloted Vehicle): Early name for UAVs, UGVs.

Sabots: Tank rounds identifiable by their 'shoes,' which initially hold the tank round in the barrel, then fall away as the round flies to its target. *See* APFSDS.

SACLOS (Semi-Automatic Command to Line of Sight): User issues partial instructions to rocket or missile while it flies, but the weapon is also partly autonomous.

SADARM (Sense and Destroy ARMour missile): aircraft-delivered missile that descends by parachute until it recognizes the auditory or visual signature of a tank: rocket motors then ignite and the SADARM attacks the tank's top.

SAPs (Special Air Projects): Cold War cover name for early UAVs, also extended to original stealth aircraft like the SR-71 Blackbird.

SCBT (Stryker Combat Brigade Team): Fast-moving group of LAVs (in this case GDLS's Stryker) where each LAV carries different cargo (troops, weapons, IT equipment, medical gear, supplies). Key to RMA ideas of lightness, speed, and interoperability.

SDBs (Small Diameter Bombs): Two-metre long 113 kg bombs with the same penetrating capability as the GBU-28 'bunker buster.'

SIGINT: Signals Intelligence: Information taken from satellites and dedicated to capturing radio, radar, and other electronic transmissions. SIGINT is further subdivided: ELINT (Electronic Intelligence), COMINT (Communications Intelligence), TELINT (Telemetry Intelligence from satellites and missiles), RADINT (Radar Intelligence). *See* MASINT, HUMINT.

Spoofing: Fooling an intelligence device or sensor into believing either that a craft or human isn't there (by stealth technology) or that the craft is bigger or smaller than it actually is.

SSBs (Small Smart Bombs): Like SDBs although more advanced and still under development; powerful, miniaturized bombs particularly useful for arming small craft like UAVs.

Stand-off weapons: Hand-held anti-armour missiles.

Stovepipes: The chain of command's vertical stack of data, largely inaccessible to other forces.

Stryker: One of a family of Light Armoured Vehicles (LAVs).

Telepresence: The act of being connected, present, and able to act by remote means using joysticks and watching real-time displays (some other variations would be telerobotics, teleoptics).

TOW (Tube-launched Optically-tracked Wire-guided missile):
More powerful and farther reaching than the Soviet RPG-7, an
infantry-borne surface-to-surface (anti-tank) or surface-to-air
(anti-aircraft) hand-held guided rocket.

TSAS (Tactile Situational Awareness System): Touch-based system
that allows user to fly, drive, shoot from, and operate robotic systems
remotely without having to look at a screen.

UAV: Unmanned (or Uninhabited) Aerial Vehicle.

UCAR: Unmanned (or Uninhabited) Combat Aerial Rotorcraft.

UCAV: Unmanned (or Uninhabited) Combat Aerial Vehicle.

UGCVs (Unmanned Ground Combat Vehicles): Weaponized robot
land vehicles that execute missions with varying degrees of autonomy, from fully guided to completely autonomous.

UGV: Unmanned (or Uninhabited) Ground Vehicle.

UUV: Unmanned (or Uninhabited) Underwater Vehicle.

Warfighting: Military combat operations in accordance with RMA
principles.

Warsaw Pact: Countries of Eastern Europe united in a defence
agreement that was authored by the Soviet Union and ended in 1989
with the collapse of that country.

Works Cited

ABC News Internet Ventures. 'New NASA Technology Breakthrough in Flying.' Web page, May 2001 [accessed 25 May 2001]. Available at http://more.abcnews.go.com/sections/wnt/dailynews/wnt_flightechnology010513.html.

Alexander, David. *Tomorrow's Soldier*. New York: Avon Books, 1999.

Amer, Kenneth B., Raymond W. Prouty, Greg Korkosz, and Doug Fouse. *Lessons Learned During the Development of the AH-64A Apache Attack Helicopter*. Santa Monica, CA: RAND, RP–105, 1992.

Associated Press. 'Prisoner Abuse Bush Order: Text of Order Signed by President Bush on Feb. 7, 2002, outlining treatment of al-Qaida and Taliban detainees.' Web page [accessed 1 October 2004]. Available at http://www.federalnewsradio.com/index.php?sid=114926&nid=78&template=section_print.

Babington, Charles, and Con Oldenburg. 'GOP Brings up Draft to Knock It Down.' Web page, Wednesday, 6 October 2004; p. A01 [accessed 5 October 2004]. Available at http://www.washingtonpost.com/wp-dyn/articles/A9479-2004Oct5.html?sub=AR.

Baker, Ellen. 'Nuclear Weapons.' In Stanley I. Kutler, ed., *Encyclopedia of the Vietnam War*, 373–4. New York: Charles Scribner's Sons, 1996.

Banerjee, Neela. 'Rebuilding Bodies, and Lives, Maimed by War.' New York Times, 16 November 2003, 1, 16.

Barry, John, Michael Hirsh, and Michael Isikoff. 'The Roots of Torture.' Web page, [accessed 17 May 2004]. Available at http://msnbc.msn.com/id/4989481/.

Bateman, Robert L. III. 'Pandora's Box.' In Bateman, ed., *Digital War*, 4–52. Novato, CA: Presidio Press, 1999.

Becker, Stephen. *Dog Tags*. New York: Berkley Publishing, 1978.

Beidler, Philip D. 'The Last Huey.' *Mythosphere*, 1, no. 1 (1998): 51–64.

– *Late Thoughts on an Old War: The Legacy of Vietnam*. Athens: University of Georgia Press, 2004.

Bicheno, Hugh. 'Total War.' In Richard Holmes, ed., *The Oxford Companion to Military History*. Oxford: Oxford University Press, 2001.

Bielitzki, Dr Joseph T. 'BAA03–02, Addendum 2, Special Focus Area: Metabolic Dominance.' Web page, February 2003 [accessed 20 February 2004]. Available at http://www.darpa.mil/dso/solicitations/baa03 –02mod2.htm.

Birdwell, Dwight W., and Keith William Nolan. *A Hundred Miles of Bad Road*. Novato, CA: Presidio Press, 1997.

Boeing Company. 'AH-64 Apache Multi-Mission Combat Helicopter.' Web page, 2001 [accessed 3 July 2001]. Available at http://www.boeing.com/ rotorcraft/military/ah64d/ah64d.htm.

Boeing Sikorsky. 'RAH-66 Comanche.' Web page, 2000 [accessed 3 July 2001]. Available at http://www.RAH66COMANCHE.com.

– 'RAH-66 Comanche Capability.' Web page, 2000 [accessed 3 July 2001]. Available at http://www.RAH66COMANCHE.com/capable.html.

Bolger, Daniel P. *Death Ground*. Novato, CA: Presidio Press, 2000 (1999).

Bonsor, Kevin. 'How Spy Flies Will Work.' Web page [accessed 23 July 2003]. Available at http://people.howstuffworks.com/spy-fly.htm.

Book, Elizabeth G. 'Competition Gets Under Way for Objective Force Warrior.' *National Defense* 86, no.582 (2002): 32–3.

Boothby, Neil G., and Christine M. Knudsen. 'Children of the Gun.' *Scientific American* 282, no.6 (2000): 60–5.

Bourke, Joanna. *An Intimate History of Killing*. London: Granta Books, 1999.

Bowden, Mark. *Black Hawk Down: A Story of Modern War*. New York: Atlantic Monthly Press, 1999.

Bowman, John S., ed. *The Vietnam War: Day by Day*. London: Bison Books, 1989.

Boyer, Peter J. 'A Different War.' *New Yorker* (1 July 2002): 54–67.

Brennan, Matthew. *Brennan's War: Vietnam, 1965–1969*. New York: Pocket Books, 1986.

– *Headhunters*. New York: Pocket Books, 1988.

Bush, G.W., President. 'President Addresses the Nation in Prime Time Press Conference.' Web page [accessed 6 June 2004]. Available at http://www.whitehouse.gov/news/releases/2004/04/20040413–20.html.

Cameron, David (staff editor). 'Artificial Muscles Gain Strength.' Web page, February 2002 [accessed 20 February 2002]. Available at http://www.technologyreview.com/articles/print-version/cameron021502.asp.

Carlock, Chuck. *Firebirds*. New York: Bantam Books, 1997.

Carmichael, Bruce W., Col. (Sel.), et al. *StrikeStar 2025: A Research Paper Presented to Air Force 2025*. U.S. Air Force, 1996.

Clancy, Tom. *Armored Cav*. New York: Berkley Books, 1994.

Clausewitz, Carl Von. *On War*. Translated and edited by Michael Howard and Peter Paret. Princeton, NJ: Princeton University Press, 1989.

Clynes, Manfred E., and Nathan S. Kline. 'Cyborgs and Space.' In Chris Hables Gray with Heidi J. Figueroa-Sarriera and Steven Mentor, eds., *The Cyborg Handbook*, 29–33. New York and London: Routledge, 1995.

Coleman, J.D. *Pleiku: The Dawn of Helicopter Warfare in Vietnam*. New York: St Martin's Press, 1988.

Committee to Review the Health Effects in Vietnam Veterans of Exposure to Herbicide. *Veterans and Agent Orange: Update 2002*. Washington, DC: National Academies Press, 2003.

Cooper, Belton Y. *Death Traps*. Novato, CA: Presidio Press, 1998.

Dean, Eric T., Jr. *Shook over Hell*. Cambridge, MA: Harvard University Press, 1997.

Defense Advanced Research Projects Agency (DARPA). 'BAA 01–42, Addendum 1, Special Focus Area: Brain Machine Interfaces.' Web page [accessed 23 August 2002]. Available at http://www.darpa.mil/baa/baa01–42mod1.htm.

– 'Continuous Assisted Performance (CAPS).' Web page [accessed 2 June 2004]. Available at http://www.darpa.mil/dso/thrust/biosci/cap.htm.

Department of the Army, United States of America. *Force XXI Operations: A Concept for the Evolution of Full-Dimensional Operations for the Strategic Army of the Early Twenty-First Century*. Fort Monroe, VA: Headquarters, United States Army Training and Doctrine Command, 1994.

– *Joint Vision 2020*. Washington, D.C.: U.S. Government Printing Office, 2000.

Dick, Philip K. 'Autofac.' *The Minority Report and Other Classic Stories.*
New York: Citadel Press, 2002.

Doleman, Edgar C., Jr, and editors of Boston Publishing. *Tools of War.* Boston: Boston Publishing, 1984.

Domenici, Senator Pete V. 'Directed Energy and Non-Lethal Use of Force.'
Congressional Record, 20 March 2001, S2572. Web page, March 2001
[accessed 9 October 2003]. Available at http://www.fas.org/sgp/congress/
2001/s032001.html.

Douglas, Mary. *Purity and Danger.* London and Boston: Routledge and
Kegan Paul, 1979 (1969).

du Picq, Colonel Ardant. *Battle Studies.* Harrisburg, Penn.: Military Service
Publishing, 1947.

DukeMedNews [Duke University]. 'Monkeys Consciously Control a Robot
Arm Using Only Brain Signals; appear to "Assimilate" Arm As If it Were
Their Own.' Web page [accessed 9 November 2004]. Available at http://
dukemednews.org/news/article.php?id=69.

Duke University, News and Communications. 'DARPA to Support Development of Human Brain-Machine Interfaces.' Web page, August 2002
[accessed 17 June 2004]. Available at http://www.dukenews.duke.edu/
news/newsreleaseb4e9.html?p=all&id=7 37&catid=2.

Edwards, Sean J.A. *Swarming on the Battlefield: Past, Present, and Future.*
Santa Monica, CA: RAND, MR–1100–OSD, 2000.

Ehrhart, W.D. *Passing Time.* Amherst: University of Massachusetts Press,
1995.

– *Vietnam-Perkasie.* New York: Zebra Books, 1985.

Ellis, John. *The Sharp End.* New York: Charles Scribner's Sons, 1980.

– *The Social History of the Machine Gun.* London: Pimlico 1993 (1976).

English, John A., and Bruce I. Gudmundsson. *On Infantry.* Westport, CT:
Praeger Publishers, 1994 (1981).

Enzensberger, Christian. *Smut: An Anatomy of Dirt (Grösserer Verusch
Über Den Schmutz).* Sandra Morris, trans. New York: Continuum, 1972
(1968).

Erwin, Sandra I. 'Army "Transformation" Plans Could Be Revisited after
War.' *National Defense* 87, no.594 (2003): 22–3.

– 'Future Howitzer Is Not "Son of Crusader".' *National Defense* 87, no.594
(2003): 22–3.

Evans, Michael. 'Close Combat: Lessons from the Cases of Albert Jacka and

Audie Murphy.' In Michael Evans and Alan Ryan, eds., *The Human Face of Warfare*, 37–53. New South Wales, Australia: Allen and Unwin, 2000.

Fahey, Dan. 'Collateral Damage: How U.S. Troops Were Exposed.' In John Catalinotto and Sara Flounders, eds., *Metal of Dishonor: Depleted Uranium*, 25–42. New York: International Action Center, 1999.

Federation of American Scientists, Military Analysis Network. 'Land Warrior.' Web page, August 1999 [accessed 3 November 2002]. Available at http://www.fas.org/man/dod-101/sys/land/land-warrior.htm.

– 'M1 Abrams Main Battle Tank.' Web page, April 2000 [accessed 5 June 2003]. Available at http://www.fas.org/man/dod-101/sys/land/m1.htm.

– 'Rand Review Indicates No Evidence of Harmful Health Effects from Depleted Uranium.' Web page, April 1999 [accessed 5 June 2003]. Available at http://www.fas.org/man/dod-101/sys/land/docs/b04151999_bt170 -99.html.

Förster, Stig, and Jörg Nagler, ed. *On the Road to Total War*. New York: Cambridge University Press, 2002.

Foucault, Michel. *Discipline and Punish: The Birth of the Prison*. Alan Sheridan, trans. New York: Vintage Books, 1979.

Frankowski, Leo. *A Boy and His Tank*. Riverdale, N.Y.: Baen Publishing, 2000.

Friedman, George and Meredith. *The Future of War*. New York: Crown Publishers, 1996.

Fritz, Stephen G. *Frontsoldaten*. Lexington: University Press of Kentucky, 1995.

Fulghum, David A. 'Dumb Blobs of Light.' *Aviation Week & Space Technology* 159, no.7 (2003): 47–8.

– 'Network Warfare: Hope and Hype.' *Aviation Week & Space Technology* 157, no.20 (2002): 33.

– 'New Bag of Tricks.' *Aviation Week & Space Technology* 158, no.16 (2003): 22–4.

– 'Predator's Progress.' *Aviation Week & Space Technology* 158, no.9 (2003): 48–50.

– 'Stealth UAV Goes to War.' *Aviation Week & Space Technology* 159, no.1 (2003): 20–1.

– 'Targets Become UAVs.' *Aviation Week & Space Technology* 159, no.4 (2003): 54–5.

- 'UAVs Whet the Appetite.' *Aviation Week & Space Technology* 158, no.9 (2003): 52.
- 'USAF Acknowledges Beam Weapon Readiness.' *Aviation Week & Space Technology* 157, no.15 (2002): 27–8.
- and Robert Wall. 'Deployment of New Technology Continues.' *Aviation Week & Space Technology* 158, no.4 (2003): 37.
- 'Small UAVs to Carry Disposable Pulse Weapons.' *Aviation Week & Space Technology* 157, no.18 (2002): 60.
- 'UAVs Validated in West Bank Fight.' *Aviation Week & Space Technology* 156, no.19 (2002): 26–7.
- 'USAF Tags X-45 UCAV as Penetrating Jammer.' *Aviation Week & Space Technology* 157, no.1 (2002): 26–7.

Fussell, Paul. *The Boy Scout Handbook and Other Observations*. New York: Oxford University Press, 1982.
- *Doing Battle*. Boston: Little, Brown, 1996.
- *The Great War and Modern Memory*. London: Oxford University Press, 1975.
- *Thank God for the Atom Bomb*. New York: Summit Books, 1988.
- *Wartime*. New York: Oxford University Press, 1989.

Gabriel, Richard A. *No More Heroes*. New York: Hill and Wang, 1987.

Garamone, Jim. 'Army Tests Land Warrior for 21st Century Soldiers.' Web page, June 2004 [accessed 17 June 2004]. Available at http://www.defenselink.mil/news/Sep1998/n09111998_9809117.html.

Gibson, James William. *The Perfect War: Technowar in Vietnam*. Boston: Atlantic Monthly Press, 1986.

Gilbert, Oscar E. *Marine Tank Battles in the Pacific*. Conshohocken, PA: Combined Publishing, 2001.

Ginzberg, Eli, et al. *The Ineffective Soldier: Lessons for Management and the Nation*. New York: Columbia University Press, 1959.

Giroir, Dr Brett P. (program manager). 'Preventing Sleep Deprivation (PSD).' Web page [accessed 5 June 2004]. Available at http://www.darpa.mil/dso/thrust/biosci/cap.htm.

Glenn, Russell W. *Reading Athena's Dance Card*. Annapolis, MD: Naval Institute Press, 2000.

Golomb, Beatrice Alexandra. *A Review of the Scientific Literature As It Pertains to Gulf War Illnesses: Volume 2 Pyridostigmine Bromide*. Santa Monica, CA: RAND National Defense Research Institute, 1999.

Gonzalez, Juan. 'Poisoned? Shocking Report Reveals Local Troops May Be Victims of America's High-Tech Weapons.' Web page, April 2004 [accessed 3 May 2004]. Available at http://www.ngwrc.org/NewsArticle.cfm?NewsID-844.

Goodman, Amy. 'U.S. Army Veteran Sues over "Back-Door Draft" Recall to Active Duty.' Web page, November 2004 [accessed 13 November 2004]. Available at http://www.democracynow.org/article.pl?sid=04/11/12/1518257.

Graham-Rowe, Duncan. 'Nuclear-powered Drone Aircraft on Drawing Board.' Web page [accessed 22 August 2003]. Available at http://www.newscientist.com/news/print.jsp?id=ns99993406.

Grange, David L., Huba Wass de Czege, Rich Liebert, Chuck Jarnot, Al Huber, and Mike Sparks. *Air-Mech-Strike*. Paducah, KY: Turner Publishing, 2002.

Grasmeyer, Joel M., and Matthew T. Keennon. 'Development of the Black Widow Micro Air Vehicle.' *American Institute of Aeronautics and Astronautics* AIAA–2001–0127 (2001).

Graves, Robert. *Good-bye to All That*. London: Penguin Books, 1960.

– and Spike Milligan. *Dear Robert, Dear Spike*. Pauline Scudamore, ed. Stroud, U.K.: Alan Sutton, 1991.

Gray, Chris Hables. 'The Perpetual Revolution in Military Affairs.' In Robert Latham, ed., *Bombs and Bandwidth*. New York: New Press, 2003, 199–217.

– *Postmodern War: The New Politics of Conflict*. New York, London: Guilford Press, 1997.

Grossman, Dave, Lt.-Col. 'The Bullet-Proof Mind.' Carrollton, TX: Calibre Press, 1999.

– *On Killing*. Boston: Little, Brown, 1995.

– 'Killology Research Group.' Web page, June 2002 [accessed June 2002]. Available at http://www.killology.com/audio.htm.

– 'Killology Research Group.' Web page, [accessed 6 June 2004]. Available at http://www.killology.com.

Guderian, Heinz. *Achtung-Panzer!* London: Cassell, 1999.

Hackworth, Col. David H. 'Oops, More Unexpected Casualties.' Web page, September 2002 [accessed 18 November 2002]. Available at http://www.alternet.org/story.html?StoryID-14169.

– 'Uncle Sam Will Soon Want Your Kids.' Web page [accessed 11 October

2004]. Available at http://www.sftt.org/cgi-bin/csNews/csNews.cgi ?database=Hacks%20Target.db&command=viewone& op=t&id=92&rnd= 503.4457051676911.

Haldeman, Joe. 'air support.' *DX*. Unpublished manuscript, 1989.

– *DX*. Unpublished manuscript version, 1989.

– *Forever Peace*. New York: Ace Books, 1998 (1997).

– *The Forever War*. New York: Avon Books, 1997 (1975).

– with Charles G. Waugh and Martin Harry Greenberg, ed. *Body Armor: 2000*. New York: Ace Science Fiction Books, 1986.

Harley, Naomi H., et al. *A Review of the Scientific Literature As It Pertains to Gulf War Illness: Volume 7 Depleted Uranium*. Santa Monica, CA: RAND National Defense Research Institute, 1999.

Hasford, Gustav. *The Phantom Blooper*. New York: Bantam, 1990.

– *The Short-Timers*. New York: Bantam Books, 1983.

Heinemann, Larry. *Close Quarters*. New York: Farrar, Straus, Giroux, 1977.

Heinlein, Robert A. *Starship Troopers*. New York: Berkley Medallion Books, 1968 (1959).

Heller, Joseph. *Catch-22*. New York, Toronto: Everyman's Library, 1995.

Herr, Michael. *Dispatches*. New York: Vintage Books, 1991.

Hersh, Seymour M. *Against All Enemies*. New York: Library of Contemporary Thought, 1998.

– *Chain of Command*. New York: HarperCollins, 2004.

Hogg, Ian. *Tank Killing*. New York: Sarpedon, 1996.

Holden, Wendy. *Shell Shock*. London: Channel 4 Books, 2001.

Ignatieff, Michael. *Virtual War: Kosovo and Beyond*. Toronto: Viking, 2000.

Kaiser Electronics. 'Helmet Mounted Displays.' Web page, 2001 [accessed 3 July 2001]. Available at http://www.kaiserelectronics.com/pages_00q1/ hm.html.

Kandebo, Stanley W., et al. *Aviation Week & Space Technology 2003 Aerospace Source Book*. New York: McGraw-Hill, 2003.

Keegan, John. *The Face of Battle*. New York: Barnes and Noble, 1993.

Lehrack, Otto J. *No Shining Armor*. Lawrence: University Press of Kansas, 1992.

Lewis Publishing. 'Agent Orange Website' [accessed 6 June 2004]. Available at http://www.lewispublishing.com/orange.htm.

Lifton, Robert Jay. *Home from the War*. New York: Simon and Schuster, 1973.

Linderman, Gerald F. *The World within War*. Cambridge, MA: Harvard University Press, 1999.

Manchester, William. *Goodbye, Darkness*. New York: Dell Publishing, 1987 (1979).

Marlowe, David H. *Psychological and Psychosocial Consequences of Combat and Deployment: With Special Emphasis on the Gulf War*. Santa Monica, CA: RAND National Defense Research Institute, 2001.

Marshall, S.L.A. *Men against Fire*. Norman: University of Oklahoma Press, 2000.

Marvicsin, Dennis J., and Jerold A. Greenfield. *Maverick: The Personal War of a Vietnam Cobra Pilot*. New York: G.P. Putnam's Sons, 1990.

Mason, Robert. *Chickenhawk*. New York: Penguin Books, 1984.

McDaid, Hugh, and David Oliver. *Robot Warriors*. London: Orion Media, 1997.

McIntyre, Jamie. 'Rare View of U.S. Strike on Taliban forces.' Web page, September 2002 [accessed 30 August 2003]. Available at http:// edition.cnn.com/2002/US/09/06/afghan.predator.video.

MEMS Exchange. 'What Is MEMS?' Web page, [accessed 21 July 2003]. Available at http://www.mems-exchange.org/MEMS/what-is.html.

MIT [Massachusetts Institute of Technology] News. 'Army Selects MIT for $50 Million Institute to Use Nanomaterials to Clothe, Equip Soldiers.' Web page, March 2002 [accessed 23 June 2002]. Available at http:// web.mit.edu/newsoffice/nr/2002/isn.html.

Moore, Michael. 'My First Thoughts after the Election ...' Web page, November 2004 [accessed 4 November 2004]. Available at http:// www.michaelmoore.com/words/message/index.php?messageDate =2004–11–04.

Mountcastle, John W. *Flame On!* Shippensburg, PA: White Mane Books, 1999.

National Research Council. *Uninhabited Air Vehicles*. Washington, DC: National Academy Press, 2000.

– Committee on Army Unmanned Ground Vehicle Technology, Board on Army Science and Technology, and Division on Engineering and Physical Sciences. *Technology Development for Army Unmanned Ground Vehicles*. Washington, DC: National Academies Press, 2003.

O'Driscoll, Patrick. 'Losing a Limb Doesn't Mean Losing Your Job.' Web

page, May 2004 [accessed 5 May 2004]. Available at http://www
.usatoday.com/news/nation/2004–05–05–cover-fit-to-serve_x.htm.

Office of the Secretary of Defense. *Unmanned Aerial Vehicles Roadmap
2002–2027*. Washington, DC: U.S. Department of Defense, 2002.

O'Hanlon, Michael. *How to Be a Cheap Hawk: The 1999 and 2000 Defense
Budgets*. Washington, DC: Brookings Institution Press, 1998.

– *Technological Change and the Future of Warfare*. Washington, DC:
Brookings Institution Press, 2000.

Owens, Admiral Bill with Ed Offley. *Lifting the Fog of War*. New York: Farrar, Straus and Giroux, 2000.

Perrett, Bryan. *Iron Fist*. London: Cassell, 1998.

Picou, Carol. H. 'Living with Gulf War Syndrome.' In John Catalinotto and
Sara Flounders, eds., *Metal of Dishonor: Depleted Uranium*, 43–9. New
York: International Action Center, 1999.

Pilger, John. 'Iraq, Depleted Uranium and Crimes against Humanity.' Web
page, March 2003 [accessed 8 May 2004]. Available at http://www
.arabmediawatch.com/modules.php?name=News&file=article& sid=129.

Puller, Lewis B., Jr. *Fortunate Son*. New York: Bantam Books, 1993.

Rhodes, Keith. *Gulf War Illnesses: DOD's Conclusions about U.S. Troops'
Exposure Cannot Be Adequately Supported*. Washington, DC: United
States General Accounting Office, 2004.

Richards, Chester W. 'Reforming the Marketplace: The Industrial Component of National Defense.' In Donald Vandergriff, ed., *Spirit, Blood, and
Treasure*, 293–340. Novato CA: Presidio Press, 2001.

Richardson, J.J. 'Depleted Uranium: The Invisible Threat.' Web page, June
1999 [accessed 5 June 2003]. Available at http://www.motherjones.com/
total_coverage/kosovo/reality_check/du .html.

Roberts, Les, Riyadh Lafta, Richard Gar, Jamal Khudhairi, and Gilbert Burn-
ham. 'Mortality before and after the 2003 Invasion of Iraq: Cluster Sample
Survey.' Web page, October 2004 [accessed 13 November 2004]. Available at http://image.thelancet.com/extras/04art10342web.pdf.

Rostker, Bernard. *Depleted Uranium: A Case Study of Good and Evil*. Santa
Monica, CA: RAND, 2002.

Royal Society. 'Royal Society Statement on Use of Depleted Uranium Munitions in Iraq.' Web page, April 2003 [accessed 5 June 2003]. Available at
http://www.royalsoc.ac.uk/du/.

Russ, Martin. *Breakout*. New York: Penguin Books, 2000.

Sack, John. *Company C*. New York: William Morrow, 1995.

Sajer, Guy. *The Forgotten Soldier*. Washington, DC: Brassey's, 2000.

Sayen, John. 'Force Structure and Unit Design.' In Donald Vandergriff, ed., *Spirit, Blood, and Treasure*, 167–98. Novato CA: Presidio Press, 2001.

Scarry, Elaine. *The Body in Pain*. New York: Oxford University Press, 1985.

Seltzer, Mark. *Bodies and Machines*. New York and London: Routledge, Chapman and Hall, 1992.

Severo, Richard, and Lewis Milford. *The Wages of War*. New York: Simon and Schuster, 1989.

Shepard's. 'Pentagon Plans Hypersonic UAV.' Web page, October 2003 [accessed 20 July 2003]. Available at http://www.uvonline.com/cgi-bin/viewt=N&r=N/1547.

Shephard, Ben. *A War of Nerves*. Cambridge, MA: Harvard University Press, 2001.

Sikorsky Aircraft. 'War Hero, 2027 A.D.' Web page, 2000 [accessed 3 July 2001]. No longer available. http://www.sikorsky.com/programs/blackhawk/warhero.html.

Sledge, E.B. *With the Old Breed*. New York: Oxford University Press, 1990.

Spalding, Richard D. *Centaur Flights*. New York: Ivy Books, 1996.

Spark, Alasdair. 'Flight Controls: The Social History of the Helicopter As a Symbol of Vietnam.' In Jeffrey Walsh and James Aulich, eds., *Vietnam Images: War and Representation*, 86–111. London: Macmillan Press, 1989.

Swofford, Anthony. *Jarhead*. New York: Scribner, 2003.

Takiff, Michael. *Brave Men, Gentle Heroes*. New York: HarperCollins, 2003.

Talbot, David. 'Super Soldiers.' *Technology Review* 105, no. 8 (2002): 44–51.

Tetlow, Stephen. 'Incorporating Human Factors in Simulation: A British Army View.' In Michael Evans and Alan Ryan, eds., *The Human Face of Warfare*, 25–36. New South Wales, Australia: Allen and Unwin, 2000.

Tiboni, Frank, and Bob Brewin. 'DoD's GIG-BE Readies for Prime Time.' Web page, September 2004 [accessed 15 November 2004]. Available at http://www.fcw.com/fcw/articles/2004/0927/news-gigbe-09-27-04.asp.

Tiron, Roxana. 'Lack of Autonomy Hampering Progress of Battlefield Robots.' *National Defense* 87, no.594 (2003): 33–5.

Trumbo, Dalton. *Johnny Got His Gun*. New York: Bantam Books, 1989.

USAF Air Force Research Laboratory. 'High Power Microwaves Fact

Sheet.' Web page, September 2002. Available at http://www.de.afrl.af.mil/Factsheets/.

USAF Scientific Advisory Board. 'New World Vistas: Air and Space Power for the 21st Century: Summary Volume.' Web page, 1995 [accessed 1 July 2003]. Available at http://www.fas.org/spp/military/docops/usaf/vistas/vistas.htm.

– 'SAF/PA 96–1204 UAV: Technologies and Combat Operations. Executive Summary.' Web page, December 1996 [accessed 14 August 2003]. Available at http://www.fas.org/man/dod-101/sys/ac/docs/ucav96/afrttech.htm.

United States Army. *Force XXI Operations*. Fort Monroe, VA: Department of the Army, TRADOC Pam 525–5, 1994.

U.S. Army Environmental Policy Institute. 'Health and Environmental Consequences of Depleted Uranium Use by the U.S. Army.' Web page, June 1994 [accessed 20 February 2004]. Available at http://www.fas.org/man/dod-101/sys/land/docs/techreport.html.

United States General Accounting Office. *Veterans' Compensation: Evidence Considered in Persian Gulf War Undiagnosed Illness Claims*. Washington, DC: United States General Accounting Office, 1996.

van Creveld, Martin L. *Technology and War*. New York: Free Press, 1989.

– *The Transformation of War*. New York: Free Press, 1991.

Vernon, Alex. *The Eyes of Orion*. Kent, OH: Kent State University Press, 1999.

Vick, Karl. 'Fallujah Strikes Herald Possible Attack.' Web page, October 2004 [accessed 17 October 2004]. Available at http://www.washingtonpost.com/wp-dyn/articles/A34612-2004Oct15.html.

– 'More Troops Suffering Severe Head Wounds.' Web page, April 2004 [accessed 29 April 2004]. Available at http://msnbc.msn.com/id/4839405.

Virilio, Paul. *Open Sky*. Julie Rose, trans. London, New York: Verso, 1997.

– *Speed and Politics*. Mark Polizzotti, trans. New York: Semiotext(e), 1986.

– and Sylvere Lotringer. *Pure War*. Mark Polizzotti, trans. New York: Semiotext(e), 1983.

Vizard, Frank. 'Extinguishing the Threat: U.S. Special Weapons May Target Iraqi Chemical and Biological Threats.' Web page, February 2003 [accessed 9 October 2003]. Available at http://www.sciam.com/article.cfm?articleID=00000CB6–18E9–1E4D-967D809EC588EEDF&pageNumber=2&catID=4.

Vonnegut, Kurt, Jr. *Slaughterhouse-Five*. New York: Dell Publishing, 1971.

Wagner, William. *Lightning Bugs and Other Reconnaissance Drones*. Fallbrook, CA: Armed Forces Journal International, 1982.

Wall, Robert. 'X-45A Flies into Turbulent Future.' *Aviation Week & Space Technology* 156, no.21 (2002): 26–7.

– and David A. Fulghum. 'The Intel Battle.' *Aviation Week & Space Technology* 158, no.19 (2003): 62–3.

Westmoreland, General William C. *A Soldier Reports*. New York: Dell Publishing, 1984.

Wilcox, Fred A. *Waiting for an Army to Die*. Cabin John, MD, and Washington, DC: Seven Locks Press, 1989.

Wilcox, Gregory. 'Maneuver Warfare: More Than a Doctrine.' In Donald Vandergriff, ed., *Spirit, Blood, and Treasure*, 125–66. Novato CA: Presidio Press, 2001.

Wright, Patrick. *Tank*. New York: Viking Penguin, 2002.

Zumbro, Ralph. *Tank Sergeant*. Novato, CA: Presidio Press, 1986.

Illustration Credits

Index

Digital Futures is a series of critical examinations of technological development and the transformation of contemporary society by technology. The concerns of the series are framed by the broader traditions of literature, humanities, politics, and the arts. Focusing on the ethical, political, and cultural implications of emergent technologies, the series looks at the future of technology through the 'digital eye' of the writer, new media artist, political theorist, social thinker, cultural historian, and humanities scholar. The series invites contributions to understanding the political and cultural context of contemporary technology and encourages ongoing creative conversations on the destiny of the wired world in all of its utopian promise and real perils.

Series Editors:
Arthur Kroker and Marilouise Kroker

Editorial Advisory Board:
Taiaiake Alfred, University of Victoria
Michael Dartnell, University of New Brunswick
Ronald Deibert, University of Toronto
Christopher Dewdney, York University
Sara Diamond, Banff Centre for the Arts
Sue Golding (Johnny de philo), University of Greenwich
Pierre Levy, University of Ottawa
Warren Magnusson, University of Victoria
Lev Manovich, University of California, San Diego
Marcos Novak, University of California, Los Angeles
John O'Neill, York University
Stephen Pfohl, Boston College
Avital Ronell, New York University
Brian Singer, York University
Sandy Stone, University of Texas, Austin
Andrew Wernick, Trent University

Books in the Series:
Arthur Kroker, *The Will to Technology and the Culture of Nihilism: Heidegger, Nietzsche, and Marx*
Neil Gerlach, *The Genetic Imaginary: DNA in the Canadian Criminal Justice System*
Michael Strangelove, *The Empire of Mind: Digital Piracy and the Anti-Capitalist Movement*
Tim Blackmore, *War X: Human Extensions in Battlespace*